C0-AYI-687

SOCIETY, BEHAVIOUR, AND CLIMATE CHANGE MITIGATION

Edited by

Eberhard Jochem
Fraunhofer Institute for Systems and Innovation Research
Karlsruhe, Germany
Centre for Energy Policy and Economics
Zürich, Switzerland

Jayant Sathaye
Lawrence Berkeley National Laboratory
Berkeley, California, USA

and

Daniel Bouille
Institute for Energy Economics
Buenes Aires, Argentina

KLUWER ACADEMIC PUBLISHERS
DORDRECHT / BOSTON / LONDON

A C.I.P. Catalogue record for this book is available from the Library of Congress.

ISBN 0-7923-6802-9

Published by Kluwer Academic Publishers,
P.O. Box 17, 3300 AA Dordrecht, The Netherlands.

Sold and distributed in North, Central and South America
by Kluwer Academic Publishers,
101 Philip Drive, Norwell, MA 02061, U.S.A.

In all other countries, sold and distributed
by Kluwer Academic Publishers,
P.O. Box 322, 3300 AH Dordrecht, The Netherlands.

Printed on acid-free paper

Printed in the Netherlands.

Contents

Preface

The Intergovernmental Panel on Climate change (IPCC) was jointly established by the World Meteorological Organization (WMO) and the United Nations Environment Programme (UNEP) to assess the scientific, technical and socio-economic information relevant for the understanding of the risk of human-induced climate change. Since its inception the IPCC has produced a series of comprehensive Assessment Reports, Special Reports and Technical Papers on the state of the understanding of causes of climate change, its potential impacts, and options for response strategies.

In 1998, Working Group III (WG III) of the ongoing Third Assessment was charged by the IPCC Plenary to assess the scientific, technical, environmental, economic and social aspects of the mitigation of climate change and a series of cross-cutting issues such as equity, development and sustainability. Its mandate was changed from a predominantly disciplinary assessment of the economic and social dimensions of climate change in the Second Assessment Report to an interdisciplinary assessment of the options to control the emissions of greenhouse gases and/or enhance their sinks.

One key issue of the IPCC Third Assessment Report (TAR) on mitigation of climate change, which has undergone an extensive review by scientists and governments, is the role of present and future behaviour of individuals, households, private and public companies, public authorities and other stakeholders.

The patterns of behaviour and underlying value systems and cultures differ substantially among world regions and societal groups. The lead authors of several chapters of WG III recognized that the social, behavioural and cultural changes involved in mitigating global climate change are poorly understood. They acknowledged the need to broaden the existing conceptual framework and decided to invite psychologists, anthropologists and other social scientists to share their perspectives on the issue of climate change mitigation.

The German Government recognized that this lack of a broader conceptual framework could be addressed, at least in part, by an IPCC Expert Meeting on social, behavioural, and cultural aspects of mitigation measures and policies. The meeting was held in March 2000 at Karlsruhe, Germany, and was organised by the Fraunhofer-Institute of Systems and Innovation Research (ISI) and the Technical Support Unit of IPCC Working Group III and was attended by about 35 participants. Its major objectives were the examination of recent findings and concepts arising from the social sciences in the context of climate mitigation options. A broad set of experts from many scientific communities and geographic regions participated giving their contributions as speakers, discussants, rapporteurs, and reviewers.

The ten papers presented at the workshop are included in this volume. They represent various disciplines from the social sciences, including economics, political science, psychology, marketing, and anthropology. We are grateful for the contributions made by all participants.

Our especial thanks to Sascha van Rooijen and the TSU for their permanent support, Renate Schmitz for organizing the IPCC expert meeting and converting the manuscript of the authors into this excellent volume; equally we would like to acknowledge Irmgard Sieb's thorough work on the layout and Anne Ray's for proof reading the contributions, of non-native authors in particular.

R. T. Watson

Robert T. Watson
Chair, IPCC

NA

Introduction

The Intergovernmental Panel on Climate Change (IPCC) noted in its Second Assessment Report that greenhouse gas concentrations in the atmosphere are increasing, and that the balance of evidence suggests a discernible human influence on global climate. Stabilizing the changing global climate will require policies, programmes, and measures to deploy new mitigation technologies and entrepreneurial opportunities. The social, behavioural and cultural changes that might be involved in mitigating global climate change, however, are poorly understood at present. The conceptual framework for mitigation assessment has been developed by energy economists and technologists during the last two decades; it has traditionally focused on the concepts of economic and technical potentials, whose attainment is prevented by barriers and market imperfections.

This rather technical or physico-chemical ('activation energy') conceptual framework does not sufficiently take into account factors such as individual motivation and personality (e.g., convenience, fun culture, social prestige), communication and decision patterns within administrations and companies (e.g., hierarchical decision processes, image building) or cultural aspects and life styles. Explanations of market imperfections are typically of a microeconomic nature, e.g., transaction costs, additional risks of impaired product quality or disruption of production, energy subsidies, or consumer utility functions. The concepts used so far tend to neglect psychological, social and cultural explanations of human behaviour and choice.

These neglected aspects need to be considered with much more emphasis in order to design effective policies, programmes and measures However, there is relatively little literature on climate change mitigation policies with a strong grounding in the social sciences, especially in countries with economies in transition and developing countries. Therefore, this book is focused on the psychological, social and cultural aspects and related conceptual frameworks of mitigation assessment. Its aim is to contribute to the Third Assessment Report of the IPCC, Working Group III by helping to broaden the conceptual basis, and hence to encourage the development of more effective mitigation policies worldwide. The book will also assist marketing experts of energy service companies, technology producers and utilities in gaining a better understanding of their customers' decisions, and provide material for relevant courses at universities and other academic and research institutions.

As usual in the IPCC process, a successful publication has depended first and for most in the co-operation of scientists and other experts worldwide. In the case of this book the participants of the expert meeting mentioned in the Annex list of the book have devoted much time and effort as authors, reviewers, discussants or rapporteurs. We are extremely grateful for their commitment to the IPCC process.

Rainer Baake
State Secretary of the Environmental Ministry, Berlin, Germany

Introduction

Incorporating Behavioural, Social, and Organizational Phenomena in the Assessment of Climate Change Mitigation Options

[illegible]

[illegible]

[illegible] perfect information without the computational decision making costs. [illegible]

[illegible]

In [illegible] these issues, this paper [illegible] models might better reflect behavioural and organizational changes, and what this would imply for the effectiveness of various policies. We examine potential modifications in the way the behaviour of firms and consumers might be represented. Our intention is to provide greater insights about the economic impact of alternative policies, especially those designed to foster technological innovation and diffusion.

The paper is organized as follows. First, we review the most typical modelling framework in climate change mitigation assessment – the Computable General Equilibrium (CGE) approach, which is based on neoclassical economic theory. Second, we revisit the major criticisms of the neoclassical model, especially in regard to

E. Jochem et al. (eds.), Society, Behaviour, and Climate Change Mitigation, 1–64.

Incorporating Behavioural, Social, and Organizational Phenomena in the Assessment of Climate Change Mitigation Options

John A. "Skip" Laitner
Stephen J. DeCanio
Irene Peters

1. Introduction

Successful climate change modelling must reflect *what people and organizations actually do.* Yet, to date, the majority of climate mitigation assessments have tended to mischaracterize the behaviour of the economic agents whose decisions ultimately affect the planet's climate. Much of this mischaracterization stems from known deficiencies in neoclassical economic theory, with its overly narrow reliance on unrealistic and unsubstantiated assumptions about the characteristics of consumers and firms.

In the neoclassical approach, businesses are presumed to maximize a well-defined profit function that arises from a highly stylized set of market and technological conditions. It is assumed that resources are utilized in an entirely efficient manner. Similarly, consumers and households are specified as maximizing a standard utility function subject only to a budget constraint. Moreover, they are assumed to have perfect information without the complications of decision-making costs. The consequence is that both consumers and businesses face a tradeoff between economic benefits and climate protection. What is seemingly a result of the neoclassical modelling exercise is already implied by the *assumptions* of that exercise.

In fact, both theory and empirical evidence provide compelling grounds for believing that environmentally related decision making processes may differ in significant aspects from the way these activities are represented in the neoclassical model. Thus, the standard numerical simulations that are based on neoclassical principles do not allow policy makers to assess the full range of options for effective and efficient carbon abatement policies.

In addition to examining the theory and evidence on these issues, this paper describes how models might better reflect behavioural and organizational change, and what this would imply for the effectiveness of various policies. We examine potential modifications in the way the behaviour of firms and consumers might be represented. Our intention is to provide greater insights about the economic impact of alternative policies, especially those designed to foster technological innovation and diffusion.

The paper is organized as follows. First, we review the most typical modelling framework in climate change mitigation assessment – the Computable General Equilibrium (CGE) approach, which is based on neoclassical economic theory. Second, we revisit the major criticisms of the neoclassical model, especially in regard to

E. Jochem et al. (eds.), Society, Behaviour, and Climate Change Mitigation, 1–64.

climate change mitigation assessments. Third, we suggest a series of short-term improvements that could begin to capture behavioural and social influences on mitigation options. Finally, we discuss longer-term research needs, closing with an epilogue that calls on all of us to invent the future rather than to await it passively.

2. The current framework

The current generation of climate change mitigation assessments is dominated by top-down models. These models tend to return moderately pessimistic estimates of the economic costs associated with reducing greenhouse gas emissions (e.g., Weyant, 1999)[1]. At the same time, a large number of bottom-up energy assessments suggest that a more optimistic outlook is possible if policies encourage cost-effective technology investments (e.g., Interlaboratory Working Group, 1997, 2000; Energy Innovations, 1997). More fundamentally, a number of writers point to the need for neoclassical theory to better reflect the human dimension within energy decisions (Stern and Aronson, 1984; Stern, 1986). Others suggest the need for a new synthesis between economics and other disciplines such as sociology and behavioural psychology (Simon, 1947 and subsequently; Etzioni and Lawrence, 1991; Williamson, 1992).

The analytical gaps between the top-down and bottom-up studies and the more fundamental theoretical gap between economics and other disciplines have engendered a "paradigm drift" in mitigation assessments. Some authors suggest that the conceptual framework of conventional models suffers from a truncated awareness of the social interactions that take place in the real world (Ravetz, 1999). Others argue that the predominant mitigation assessment models offer only "a mixture of conditionally-valid, uncertain and indeterminate knowledge – data, theory and insights – combined together in a fashion which generates further indeterminacy" (Shackley and Darier, 1998). These deficiencies, in turn, foster a less-than-satisfying discussion about potential climate mitigation strategies. We propose that models offer better insights when social, behavioural, and organizational phenomena are more fully captured in climate mitigation assessments.

2.1. *DESCRIPTION OF CURRENT FRAMEWORK*

The predominant models now used to evaluate climate change mitigation policies are macroeconomic models that depict the economy as a whole. Based on national accounting data, these models record the cost of production within individual sectors (including payrolls and other expenses), the supply and consumption of goods and services, the level of investments, government taxation and transfers, and foreign trade.

---

[1] It should be kept in mind that consistent and correct application of basic economic principles suggests that it is reasonable to begin reducing emissions now because of the large potential damages from climate change and the risks of triggering an irreversible climate catastrophe (Cline, 1992; DeCanio, 1997).

Some of the macroeconomic models used in climate change policy analysis are econometrically based; sometimes they take account of the business cycle; and they may incorporate a detailed technological linear programming module on energy generation and use[2]. However, since the 1980s there has been a trend towards the application of Computable General Equilibrium models, and these now are the primary tool used for policy analysis[3]. We focus our review on the kinds of applied CGE models currently in use, although several of our criticisms apply to other modelling approaches as well[4].

The trend towards using CGE models in applied policy analysis mirrors, albeit with a time lag, a strong movement in academic economics to put macroeconomics on "microfoundations". By this we mean the program to derive the behaviour of economic aggregates in terms of the microeconomic theory of individual firms and consumers. Applying neoclassical microeconomic theory to an economy-wide context (this is the essence of General Equilibrium Theory) has severe difficulties which cast doubt on the usefulness of CGE models for policy analysis. The difficulties are well known and discussed in the technical economic literature. In fact, it is in the specialized literatures dealing with the theory of the firm, welfare economics, aggregation, and the behavioural roots of economic activity, that the modelling issues and criticisms we discuss here have been developed[5]. However, *these theoretical advances have not yet been widely assimilated into the applied CGE models that are being used for policy analysis, and for climate policy analysis in particular.*

CGE models are said to have "microfoundations" in that the mathematical relationships they contain are derived from, or are consistent with, the neoclassical economic theory of producer and consumer behaviour. This theory stipulates that economic agents optimize under constraints: consumers choose their mix of consumption and, in some settings, "leisure" so as to maximize utility, given a budget constraint. Producers maximize profits, given a production technology. These optimization exercises give rise to demand and supply functions that relate quantities demanded and supplied to prices. Consumers are described by commodity demand and labor supply functions, and producers are described by input (including labor) demand and output supply functions.

---

2 Alan Manne's and Richard Richels' "ETA-Macro" combines an aggregate CGE model for the whole economy with MARKAL, a detailed linear programming energy model (Manne and Richels, 1992).

3 The range of models discussed in Weyant et al. (1999) reflects this point. Regarding the use of the term "macro" in conjunction with CGE models, see Section 2.2.7.

4 Of course we can describe these models only in the most cursory form, highlighting features relevant for our discussion. For a survey on the state of the art in CGE modelling, see Ginsburgh and Keyzer (1997).

5 Kirman (1998) observes, "[w]hether future and, hopefully, better descriptions of economic activity are built on the foundations provided by the analysis developed here," – that is, General Equilibrium Theory – "or whether they emerge from a very different paradigm, remains to be seen....The fact that [modern economic theory] has severe limitations, and that these are clearly understood, is, paradoxically, a tribute, in itself, to the rigor of the analysis..." (p. 7 f.).

While the neoclassical economic theory of consumer and producer behaviour is formulated for the *individual* consuming and producing agents (hence, this is called "microeconomics" in the jargon of the discipline), in CGE models the theory is used to represent the behaviour of *aggregates* of these groups. These models employ the "representative agent" device, describing production in each sector with a "representative producer" and consumption with a "representative consumer" (or, depending on the level of model detail, several such representative consumers for population groups differing by a characteristic like income or location). These representative agents do not interact except indirectly in that the functions characterizing them contain the same endogenous price variables. We elaborate on this point in Section 2.2.7. below.

CGE models also assume that markets clear – that is, the quantity supplied equals the quantity demanded in each market. This constitutes "equilibrium." The models are called "*general* equilibrium models" because equilibrium is achieved in all markets simultaneously, and markets are interrelated through prices and other variables. For example, the demand function for cars contains variables for the price of cars, the prices of other goods and services (such as gasoline, airfares, etc.), and consumer income. This captures the interactions between different markets.

In sum, a CGE model consists of a system of equations with algebraic functions that relate demand and supply quantities to prices, as well as conditions that ensure clearance in all markets. Following the theory that underlies these models (General Equilibrium Theory), a number of assumptions are typically made about the shape of the algebraic functions. These assumptions ensure that the system of equations that make up the model has a solution, and that the solution is unique.

CGE models (and the underlying General Equilibrium Theory) do not depict how equilibrium is achieved, i.e., they do not deal with how economic agents set prices and react to them. Equilibrium is simply *assumed;* solving a CGE model is equivalent to solving a simultaneous equation system. Since abstracting from the process of price setting and adjustment is not very satisfactory, the theory has recourse to the device of the "hypothetical auctioneer", which was invented by the pioneer of general equilibrium analysis, Leon Walras. The hypothetical auctioneer quotes prices to the economic agents, observes the quantity quotations with which they respond, and adjusts prices until he has found the set of equilibrium prices at which all markets clear simultaneously. But even this artificial "tâtonnement" ("groping") process is only assumed and not explicitly modeled in CGE analysis. In fact, if it were, it would not necessarily converge to equilibrium. Again, we return to this point later in Section 2.2.7.

Dynamics can enter these models in a variety of ways. Most models have producers with durable capital stocks that are updated each period through depreciation and investment. Some models assume forward-looking agents that hold expectations over future variable values and optimize over time. Exogenous population growth based on demographics or exogenous shifts of the production function that reflect technical progress can also bring dynamics to the model. Thus, CGE models do contain some dynamics, but the dynamics are lacking in essential respects. Neither disequilibrium nor movement towards or away from equilibrium is incorporated in the typical CGE model. The tâtonnement process as described above takes place anew within each time period, using the variables that have changed over time (shifts in the production function, population size) as given. This reduces the dynamics to a shift in equilibria.

The parameters of a CGE model are usually determined through a mixture of calibration and econometric estimation. The model is calibrated to a benchmark dataset, assuming the dataset represents the economy in equilibrium (i.e. with supply equaling demand in all markets). Ordinarily, there are far more parameters than can be identified from the data; hence the parameters are either estimated piecewise, are based on econometric estimates reported in the literature, or are simply assumed[6]. As with forecasting and simulation models in general, the outcomes of CGE models are sensitive to parameter specification with the consequence that different results can be generated through different choices of parameters (Repetto and Austin, 1997).

Typical policy analysis with CGE models consists of perturbing the model with variations in policy parameters (e.g., tax rates) and computing the resulting new "counterfactual" equilibrium. This type of exercise, very common in economic analysis, is called "comparative statics".

2.2. *A GENERAL CRITIQUE OF THE CURRENT FRAMEWORK*

CGE models can be criticized at several levels. First, traditional CGE models employ a number of assumptions about agent behaviour and societal institutions that obviously lack realism. Hence, they are quite likely to produce unrealistic outcomes[7]. The most fundamental problematic assumption is that of the perfectly rational actor who optimizes an objective function under well-defined constraints. Evidence is strong that real-world actors behave very differently. Their rationality is "bounded"; they are embedded in a social and technological context that provides them with limited information and gives shape to their ambitions and desires. Firms, in particular, are social systems made up of interacting people rather than mere automatons that only compute profit-maximizing combinations of production inputs.

Yet another unrealistic assumption concerns the shape of technology. Policy analysis models typically rule out increasing returns to scale that are, in fact, quite common in production and consumption systems. However, increasing returns have

6 Shoven and Whalley (1992) observe, "...most of the applied general equilibrium models are not tested in any meaningful statistical sense...there is no statistical test of the model specification....With enough flexibility in choosing the form of the deterministic model, one can always choose a model so as to fit the data exactly. Econometricians, who are more accustomed to thinking in terms of models whose economic structure is simple but whose statistical structure is complex (rather than vice versa), frequently find this a source of discomfort" (p. 6).

7 Milton Friedman's "as if" argument – the idea that unrealistic assumptions can be warranted if they lead to correct prognoses – is discussed in Section 2.2.4.

important implications for the way in which technological change unfolds. Hence, models that omit increasing returns cannot deal with this topic very well.

The second level of critique applies to the normative implications of CGE analysis. Given that all the problematic assumptions hold, model outcomes are said to be "efficient". As we shall discuss later, this property breaks down as the many assumptions inevitably are violated.

The third level of critique has to do with gaps in the theoretical structure of CGE analysis itself, even if the unrealistic assumptions underlying it were to hold. Mathematical representations of the micro units making up the economy as rational maximizers do not translate into the same kinds of equations and functional forms at the aggregate level. More seriously, the absence of a rigorously grounded specification of the dynamics of the economy means that the solutions of the CGE models may not correspond to any actual behaviour or outcomes in the real economy.

Finally, the concept of "welfare" that these models employ is far too narrow. The models do not capture many things that people may deeply care about because these things are not traded in a market and do not command a price. This approach also prevents a number of promising "policy handles" from being recognized in the models. In addition, distributional issues are usually ignored (when they arise between living populations) or treated in a manner that is ethically hard to defend (when they arise between the living and future generations – the latter's welfare is usually discounted, thus giving it less weight than the current generation's.).

In what follows, we attempt to provide a perspective on these issues. Although an exhaustive survey is beyond the scope of this paper, we hope that our review presents a basis for assessing the current practice in climate policy modelling. Also, we provide references to the literature for the interested reader who wishes to pursue the various topics we discuss.

We begin with a critique of the normative content of models, because this point has immediate policy implications. (Section 2.2.1). Then we discuss the assumptions pertaining to missing markets (Section 2.2.2) and the shape of technology (Section 2.2.3). Sections 2.2.4 to 2.2.6 tackle the big issue of the rational actor paradigm, as it is applied to the behaviour of organizations and individuals. Section 2.2.7 moves on to the inherent theoretical difficulties of the models, which would be present even if all the critical assumptions discussed earlier were to hold. Section 2.2.8 concludes with a reflection on the erroneous concept of "welfare" that the models employ, and the importance of distributional considerations.

2.2.1. *Normative content*

CGE models that embrace neoclassical microfoundations, with their equations based on utility and profit maximization, have an important property: given a number of assumptions typically made in policy applications, the general equilibrium solution to the model is unique and constitutes an "efficient" resource allocation[8]. "Efficiency" here is understood as "Pareto-efficiency", implying that no change can be made improving the situation of any one agent without worsening that of another.

In this setting, government intervention has a welfare cost as it pulls resources away from their supposedly most productive uses. Taxes on labor income, for example, drive a wedge between the marginal benefit from work (the output it produces) and its marginal cost (the disutility arising to the worker); commodity taxes drive a wedge between the marginal cost of producing a commodity and the marginal utility of consuming it. It is the equality of these marginal rates that makes a resource allocation "efficient".

That government intervention in the marketplace should *by definition* cause distortions and welfare losses is a strong normative conclusion[9]. Even though it pertains to a highly stylized model world, this conclusion has seeped into political rhetoric and is treated as an established fact, rather than as the consequence of modelling assumptions. The perspective that government activity disturbs the otherwise efficient working of the market forces pervades the debate on the role of government in a market-based society. Yet, the theoretical result only holds if the underlying model assumptions are valid. What does this imply for the efficiency of real-world market economies? There is no doubt that competitive market-based economies excel at coordinating myriad activities and do an impressive job at mobilizing societal resources. But concluding that free markets bring about the best arrangement of production and consumption *per se*, and then concluding that government intervention necessarily has a welfare cost, is not warranted. As we pointed out before, the problem is not that theoretical work has failed to explore the assumptions that are necessary to ensure the existence, uniqueness, and efficiency of CGE model solutions. Theorists have devoted a great deal of effort to examining how to weaken the strong assumptions on which the theory must rely. Rather, the difficulty lies in CGE *applications*, in which those assumptions are used, unquestioned, either for convenience, out of habit, or in order to arrive at a particular policy conclusion.

2.2.2. *Externalities and missing markets*

The notion of "externality" is widely invoked within and outside of economics. It has become part of the political lexicon. In economic theory, the concept of an externality is used to explain the problem of environmental pollution. An externality is present if the social and the private cost of an activity diverge – i.e., if an economic agent (producer or consumer) does not bear the full social cost of his actions. (The difference between the social and the private cost is the "external cost"). The cyanide wastes of the Australian-Romanian mining firm which devastated fish stocks and ecosystems in the Tisza and Danube rivers in spring 2000 constitute an external cost because the mining firm does not have to incorporate these costs (health damage, loss of livelihood for fishermen, etc.) in its balance sheet.

---

8 "Resource allocation" is the technical term economists use to denote the arrangement of production and consumption.

9 See also Shoven and Whalley (1992), p. 34.

Economics also recognizes positive externalities, or "external benefits". The pollinating activities of bees are a classic textbook example of a positive externality that is also highly topical in Western Europe, as the decline in the stock of bees is threatening the survival of flora in many regions. The pollinating bees not only produce benefits for the beekeeper, but for the farmer whose orchard is pollinated, and for all those who derive benefits from a diverse and intact flora. The farmer and the other beneficiaries of the pollinating bees usually do not compensate the beekeeper for this "service".

In the neoclassical model, externalities constitute "distortions" (like the tax on labor discussed in Section 2.2.1). Their presence implies that general equilibrium is not Pareto-efficient, because Pareto-efficiency relies on the social cost of an activity being equal to its private cost. A long tradition in economics advocates "internalizing" externalities by equalizing private and social costs, in order to restore efficiency. In the 1920s, the economist A. C. Pigou suggested that a tax be levied on the offending activity in the amount of the marginal external cost, in order to restore market efficiency. An immense literature on "Pigouvian taxes" ensued, and the concept of internalizing externalities has become part of the rhetoric of environmental policymaking.

In fact, economic theory explains, or circumscribes, the presence of externalities with the notion of "missing markets". If there were markets for all the impacts that constitute external costs or benefits – i.e., a market for clean air, clean water, quiet, etc. – there would be no externalities[10]. It is then the task of legislation and policy to create prices or markets for these environmental goods.

There is no doubt that environmental taxes, like price signals and incentive-based mechanisms in general, can be a very effective means to affect the actions of consumers and firms and provide stimuli for targeted technological innovations. Yet, while they can make a significant contribution towards alleviating environmental problems, Pigouvian taxes do not necessarily fix all the problems, as seems to be suggested by the call for "full internalization". Several reasons account for this. First, it is virtually impossible to assess the true size of an external cost, as there is great uncertainty about the shape of damage functions. (The market cannot help uncover these social cost functions, as it might do in the case of private cost, because by definition it does not extend to externalities.) This issue has engendered a broad literature on "Prices vs. Quantities" (Weitzman, 1974), suggesting that in cases where there is the danger that damage costs increase sharply with the amount of pollution, it is advisable to regulate, i.e., set quantity standards for emissions, rather than impose a tax whose effect on emission quantities is not known *a priori*.

---

[10] Coase (1960) has argued that even in the absence of policy intervention, groups of affected people could, if the transactions costs were low enough, get together and negotiate a payment with the polluter that would compensate him for reducing the polluting activity. Alternatively, the polluter could compensate those affected by the pollution. Either arrangement would restore efficiency, but with different consequences for the distribution of wealth depending on how the "property rights" in the environmental goods (or pollution) are assigned.

Second, only when there is a *single* externality will the efficiency-restoring tax equal the marginal external cost. As soon as there are several externalities or other distortions, the tax has to be adjusted to account for interactions among these distortions. This is an instance of the problem of "second best". (Lancaster and Lipsey, 1957; Bohm, 1967). The more externalities and other distortions there are, the more complicated the formula for internalization becomes; in fact, it soon becomes analytically intractable. Because externalities and other distortions pervade our world, there is no hope of finding the right formulas for restoring Pareto-efficiency[11]. Furthermore, what is it that should count as a "distortion"? For a whole range of public services, like insurance against unemployment and congenital health defects, it would be difficult to devise markets in which private suppliers could survive (Barr, 1992). Thus, the government activity of providing these services, financed by taxes, is part of the efficient allocation itself, rather than constituting a "distortion". (See also McKee and West, 1981). Thus, not only is the notion of "efficiency" elusive, but that of "distortion" is, too.

The point here is not to argue against taxes and other incentive-based mechanisms in climate change policy. Taxes and price signals are, in fact, powerful instruments because they transmit information in decentralized fashion[12]. Rather, the difficulty of implementing optimal tax rates suggests that other policies, perhaps in conjunction with taxes, may be appropriate to correct environmental externalities.

CGE modelers typically rule out externalities or allow for just one externality, pertaining to the problem under study (e.g., carbon emissions). Thus, when discussing the effects of a policy like taxes on "efficiency", they neglect the fact that because of pervasive externalities, the whole notion of "efficiency" (at least economy-wide Pareto-efficiency) is elusive[13].

The issue of externalities throws open a much broader question. It may not be appropriate to apply pricing methods to all goods or characteristics of our society, for a whole range of reasons[14]. The satisfaction of preferences, as expressed by demand for tradable commodities, is not the beginning and end of people's happiness. Amartya Sen maintains that it is not commodities, but the human capabilities that

---

11 For a more complete discussion of pervasive externalities, see Daly and Cobb, 1989.

12 Colander, in his speculation about what "New Millennium Economics" (the economics of the first decades of the 21st century) might look like, puts it thus: "[E]conomists still believe price incentives are important and that markets solve coordination problems, but that belief is not held with the almost religious conviction with which it was held in the neoclassical era." (2000, p. 126).

13 The work by Larry Goulder and collaborators is somewhat of an exception (Goulder, 1995; Bovenberg and Goulder, 1996, 1997; Goulder and Schneider, 1999); they have studied the interaction between different distortions in a neoclassical general equilibrium context. But while addressing the problem of several simultaneous distortions, they still restrict the distortions to a few, so that the problem remains analytically tractable.

14 We cannot survey the relevant literature here. Excellent entry points to the writings by economists and philosophers on this issue are Ackerman, Kiron, et al. (eds., 1997), and VanDeVeer and Pierce (eds., 1998). Anderson (1993) and Sagoff (1981, 1988, 1994) are among the most prominent authors. Vatn and Bromley (1994) is an influential piece by economists.

help people function in a society, which are at the core of well-being[15]. Also, people may wish to pursue their goals and ambitions in a setting other than a market. Actual choices reveal not only preferences for different goods and services, but are guided by responsibilities and commitments as well as markets and prices. Thus, it is not appropriate to subsume political or ethical choices solely within a monetary framework. The contingent valuation approach, which attempts to uncover the social value of environmental and other unpriced goods by soliciting people to participate in questionnaires, is flawed for similar and additional reasons (Vatn and Bromley, 1994)[16].

To summarize, many crucially important choices of our society cannot adequately be coordinated by the price system. To reduce all the problems associated with these many facets of life to instances of "missing markets" misses their essence. Even some of the traditional public goods such as basic education or law enforcement have moral dimensions that cannot entirely be valued quantitatively. These kinds of public goods constitute critical, if not central, influences and constraints on human behaviour. The standard argument that the "non-market" aspects of life are presumed to be exogenous and non-changing for the purpose of the analysis with CGE models is not convincing. Economically relevant behaviour is guided by more than the prices of those things easily priced. Hence, models that do not incorporate such influences misrepresent the consequences and the potential for policies. They also misjudge the welfare changes that result from them.

2.2.3. Increasing returns to scale and the nature of technological change

"Increasing returns to scale" in production is a common and important real-world phenomenon, yet CGE modelers routinely exclude it from their analysis. They instead assume "constant returns to scale". As we will explain, acknowledging the presence of increasing returns in modelling and policy is important for two reasons. First, increasing returns destroy the Pareto-efficiency of model outcomes. Second, the presence of increasing returns has deep implications for the nature of technological change, and the ability of non-market mechanisms to affect its course.

Constant returns to scale is a property of the production technology implying that the cost-minimizing input mix grows proportionally with output. Doubling the output requires a doubling of all inputs while tripling the output implies a tripling of all inputs, and so on. This means that average cost – total cost divided by total output, also called unit cost – is constant over the entire output range. By contrast, increasing returns to scale implies that average cost decreases with output. A doubling of output can be achieved by a less-than-double increase of all inputs.

---

15 See, for example, Sen (1985).

16 For example, in deriving prices for environmental goods from hypothetical valuation, important information about the environmental goods is lost. Also, the value of one environmental good may depend on the levels of other goods that are available; valuing it in isolation would miss essential aspect of its character.

With some exceptions, CGE modelers tend to rule out increasing returns because this feature complicates the model and has side-effects that might be viewed as undesirable[17]. In the presence of increasing returns, the neoclassical marginal pricing rule, which ensures Pareto-optimality, is no longer practicable. Firms obeying this pricing rule would incur losses, which would have to be covered by some kind of regulatory arrangement or subsidy. Introducing different pricing rules that ensure survival of firms (for example, average cost-plus pricing) would make the model solution Pareto-*in*efficient (see, e.g., Villar, 1996).

The assumption of constant returns to scale in CGE analysis is usually defended with the argument that competition drives firms to all operate at the same minimum cost, where they can just survive (they are said to make "normal" profits). Any firm having a higher unit cost than the others must go out of business, because competition does not allow any firm to charge a higher price than its competitors. In such a setting, the industry as a whole can be argued to exhibit constant returns to scale, as it consists of many individual firms with the same unit cost.

It is indeed plausible that any firm should have an output range over which returns are increasing, and a range where they are decreasing, implying that there is a most efficient firm size, or range of efficient sizes. The interesting question is how large this efficient range might be in comparison to the potential demand for the product. If it is small, then the stylized model of many firms operating in a competitive market that forces each to accept the going price holds some plausibility. If, however, the range of efficient firm size is large compared to market demand (that is, if there are increasing returns over a relatively wide output range), there is no room for many firms. Industries with increasing returns, in which only a few suppliers can survive, still may enjoy some competition although this competition is said to be "imperfect", meaning that firms have the power to affect the sales price of their products. In the neoclassical model, this "market power" leads to a violation of the neoclassical supply pricing rule which says that output price should equal marginal cost of production. Thus, market power causes the resulting resource allocation to be inefficient.

Imperfect competition is by far more typical of today's real-world markets than the "perfect competition" assumed in most CGE models. As we have noted, imperfect competition does not mean the absence of all competition. It implies that firms compete in a number of dimensions other than price, such as product quality, product image, customer service, and innovation. This type of market structure may spur technical progress, and some kinds of technological change may even result from the quest for market power (see, for example, Schumpeter, 1942).

---

17 The international trade literature has recognized that increasing returns is an important phenomenon (Krugman, 1979, 1980; Helpman and Krugman, 1985). For CGE models of international trade with increasing returns, see Harris (1984) and Cox and Harris (1985), and the discussion in Kehoe and Kehoe (1995). Ginsburgh and Keyzer (1997) also have a chapter on CGE models with increasing returns.

One can distinguish three forms of increasing returns to scale: those related to firm size (which we described above); "learning by doing", which is the decrease in cost associated with increasing experience in production and use; and "network externalities", related to the interaction of producers and users of a product[18].

The first form of increasing returns, related to size of operations, is typical for industries that rely on an expensive infrastructure to deliver their services, like the traditional utilities – urban water services and the transmission and distribution of electricity. It would not make sense for many firms to maintain independent distribution networks. Power transmission lines are an instance of a "natural monopoly". The same is true for transportation that is bound to a network of rails, roads or air and water corridors.

The second form of increasing returns, learning by doing, is very pronounced in industries that produce a technically complex product such as power plants or medical equipment, and in industries with a high research and development (R&D) requirement, like pharmaceuticals and complex electronics. Learning by doing implies that the more experience a manufacturer has with producing a product, and the more experience the customer has with using it, the cheaper it becomes. Technologies mature over time, as design flaws are ironed out and as protocols are developed for the financing, siting, permitting, and customizing of technologies to meet users' needs. Thus, learning by doing is really "learning by using" as much as "learning by producing". Technological learning has long been recognized in established industries – it was first "discovered" in the manufacture of airframes[19] – but is also strongly at work in renewable energies, as the dramatically declining cost for photovoltaics and wind power demonstrates.

Finally, there are "network externalities", implying that a technology becomes cheaper the more users it has, because it relies on some kind of physical or knowledge infrastructure. Fuel cell powered cars are a potential example because they may require unique fuels that are available only from specially equipped pumping stations. As the number of fueling stations and related infrastructure grows, costs are expected to decline. Another example is computer software. An important determinant of its attractiveness to the customer is the number of people who use it, because this facilitates the exchange of electronic documents.

The presence of increasing returns has profound implications for the understanding and modelling of technical change. Technologies that have a head start over others can benefit sooner from a regime of decreasing costs, making it difficult for alternatives to catch up. The positive feedback loops of increasing returns can amplify small historical accidents and thus give a push to a certain technology that enjoys a (perhaps random) initial advantage. This is one cause of "path dependence", which

---

[18] Our discussion of the different forms of increasing returns and their implications for technological change draws on Peters, Ackerman, and Bernow (1999).

[19] See Alchian (1959). Arrow (1962) cites Wright (1936) as an early observer of the phenomenon, and Asher (1956) for a survey for airframes.

implies that there is not a single path for technological development, but many possible paths. Which path will ultimately prevail can depend on small but decisive events, whether accidental or purposefully brought about by policy intervention.

Path dependence can also lead to technological "lock-in": If a lot of resources have already been spent on a technology, and much learning has gone into it, the technology is likely to be cheaper than competing alternatives which would have to be developed all anew. It is possible that inferior technologies can come to dominate the market at the expense of better alternatives. This mechanism has been at work in the nuclear power industry. In the U.S. of the mid 1950s, a number of proposals for nuclear reactor design existed, and one (the light water design) was chosen in part because precursors had already been used in nuclear submarines. There is reason to believe that other designs would have been superior in the long run (Cowan, 1990).

Recent developments in the theory of "endogenous technological change" have begun to address some of the real-world phenomena relevant to the issue. The theory has identified public good characteristics of variables that are important for technological change[20]. Human capital has been recognized as an important factor in technological change; this may call for a more dedicated effort of the public sector towards education. Likewise, R&D displays some positive externalities. While a great deal of research and invention takes place within private sector firms in response to perceived economic incentives, the benefits of this innovative activity cannot entirely be captured by the firms because of knowledge "spillovers" to other parts of the economy. This also is a type of externality that provides an opportunity for positive policy intervention. The empirical work of Jones and Williams (1998) suggests that the optimal share of GDP allocated to R&D in the U.S. economy is between two and four times higher than its actual share.

While this literature holds some promising insights, it has some distance yet to go in giving up the notion of a unique, optimal path for the unfolding of technological change that the market will uncover – provided prices are "corrected" for externalities through the use of taxes and subsidies. In this setting, the market, if not disturbed by government intervention, allocates the resources that generate technological progress (human capital, R&D, venture capital) in a manner that will achieve the highest possible pay-off. But such a notion of technological change does not do full justice to the phenomenon of path dependence. Economic models that are to be useful for energy policies must be able to accommodate positive feedbacks and the socio-economic context that affects the nature, direction and rapidity of technological change (Martino, 1999).

Indeed, the goal should be to develop models that provide a range of alternative outcomes, given different initial and interim conditions. While some outcomes would be more attractive than others, the models would offer no single metric for evaluating the alternatives. Such an approach would constitute real progress, because it would acknowledge that the choice of which technologies to pursue is up to

---

20 See Romer (1986,1990) and Lucas (1988) for pioneering papers. For applications to climate policy, see Goulder and Schneider (1999) and Sanstad (1999).

society. There is a range of possibilities and economic theory cannot tell us which is the best. That choice has to be made drawing on criteria outside the realm of economic analysis.

2.2.4. *Bounded rationality in individuals and organizations: Is the "as if" argument valid?*

The most sweeping critique of the plausibility of CGE model assumptions is the one directed at the "rational actor" paradigm. This assumption is at the core of neoclassical economic theory. It stipulates that individuals have consistent, well-ordered preferences, which amounts to the existence of an ordinal utility function. It assumes that all agents possess complete information and that, in the presence of risk, they maximize expected utility[21]. Preferences are exogenous in this framework; they are a given and are not explained. Likewise, it is assumed that firms optimize a well-defined objective function, given resources and market constraints. They are always cost-minimizing and employ their resources with complete efficiency.

Clearly, these strong assumptions are at odds with reality. People do not and cannot possess perfect information. They are limited in their cognitive abilities, they are creatures of habit, and their choices are not always consistent. Psychologists and social scientists other than economists have long known this. Studies of risk perceptions have uncovered that people regularly misjudge probabilities. Prospect theory suggests that people suffer more from losses than they enjoy corresponding gains[22]. While there have always been some economists who did not adhere to the rational actor paradigm, their influence within the economics discipline has been limited. Herbert Simon (psychologist, computer scientist, and economist in one) is a prime example. He pioneered the concept of "bounded rationality" which takes account of the fact that people have limited computational capabilities and frequently resort to rule-based behaviour to reduce complexity[23].

Things may be changing, however. It seems that the economics profession increasingly recognizes the need for a re-orientation of economic theory. For example, a recent issue of the *Journal of Economic Perspectives* contains a symposium "Forecasts for the Future of Economics", in which Richard Thaler of the University of Chicago speculates that future economics will model its subjects as less than perfectly rational, and will put a greater emphasis on learning (Vol. 14, No. 1, Winter 2000; Thaler, 2000). David Colander suggests that the "neoclassical era" might be coming to an end, and that economists will be more willing to learn from empirical observation (Colander, 2000). A recently published Special Section of the *Journal of Economic Behaviour and Organization* (Vol. 41, No. 3, March 2000) devoted to "Psychological Aspects of Economic Behaviour" carried several papers reporting

---

[21] For an extensive discussion of the limitations of expected utility theory, see Starmer (2000).

[22] The original paper on prospect theory is Kahnemann and Tversky (1979). An influential early collection of this type of work is Kahnemann, Slovic, and Tversky (1982).

[23] Conlisk (1996) provides a comprehensive survey of the literature on bounded rationality. Tisdell (1996) offers a lucid account of different concepts of rationality in economics and discusses what they imply for our understanding of economic processes.

experiments illustrating how individual behaviour is conditioned by a wider range of influences than those ordinarily considered in purely "economic" modelling[24]. A session of the January, 2000, annual meeting of the American Economic Association was devoted to "Preferences, Behaviour, and Welfare", and included papers titled "Emotions in Economic Theory and Economic Behaviour" (Loewenstein, 2000), "A Boundedly Rational Decision Algorithm" (Gabaix and Laibson, 2000), and "Thinking and Feeling" (Romer, 2000). Just how far this development has come along may be illustrated by the fact that *The Economist* recently devoted an entire section in its Christmas issue on how alternative concepts of rationality are beginning to be embraced by economists[25].

Just as real-world individuals fail to fit the rational-actor mold, so do organizations. Bounded rationality is to be expected in firms because of institutional barriers to complete optimization (DeCanio, 1993), the complexity of the computational tasks that have to be carried out (DeCanio, 1999), and the limited information-processing capability that is particularly relevant in organizations (Simon, 1997)[26]. In fact, the evidence is overwhelming that organizations, be they private or public entities, are concerned with much more than profit. They have additional concerns and constraints that make perfect profit maximization impossible. There is an extensive literature in economics and operations research devoted to relative productivity measurement, in which the performance of firms or organizations is compared to industry best practices. The bibliography in Cooper et al. (1999) contains over 1,500 references[27]. Also, if firms were uniformly efficient, it is difficult to see why there would be an active market for corporate control, or why the demand for specialized business education and for management self-help books would be so extensive (DeCanio, 1994).

Clearly, the neoclassical model only provides a "caricature of real firms" (Simon, 1997, p. 51). Nelson notes, for example, that "there is a substantial body of literature which does not see a firm simply as a profit seeking 'chooser' of inputs and technology operating within a framework of widely available technological knowledge, and known factor prices. Rather, it sees a firm as a 'social system'" (1996, p. 18). And what happens inside firms and other organizations is not something that can be ignored because it is a minor detail. Rather, in the real world, "most activities take place inside organizations, even though these organizations are linked by a network of markets". (Simon, 1997, p. 51). Thus, the focus on markets, at the expense of what goes on inside organizations, is an instance of misplaced concreteness.

---

24 The individual papers in this issue are Fan (2000), Cameron (2000), Horowitz and McConnell (2000), Humphrey (2000), Zizzo et al. (2000), Butler (2000), and Croson (2000).

25 "Rethinking Thinking," which reviews the forthcoming book *Choices, Values, and Frames,* by Kahnemann and Tversky (eds.).

26 A review of some of the material on experimental decision theory is given by Rubinstein (1998).

27 This bibliography focuses on a specific branch of this literature that uses a non-parametric linear programming approach called Data Envelopment Analysis (Charnes et al., 1978). There is also a substantial empirical literature using other statistical methods.

Psychological models of human behaviour do not require maximization of profits or "utility". Indeed, one of the main points of psychology is to explore the complexity and variety of the human mind. Of course, it is possible to make the mistake of turning "utility maximization" into a tautology, so that any behaviour, no matter how irrational or perverse, can be interpreted as the maximization of *some* utility function. But that is not science; if the study of human behaviour is to have scientific content, it must be formulated in terms of hypotheses that are falsifiable. So, models of behaviour that incorporate the importance of kinship ties, social and ethical norms, passions and emotions, or genetically induced predispositions can be expected to exhibit deviations from the maximization framework. This is indeed what is found when psychologists investigate the behaviour of people in making energy-efficiency investment decisions (Stern and Gardner, 1981; Dennis et al., 1990; Stern, 1992).

Yet CGE modelers are reluctant to give up the rational actor paradigm. This has deeper causes. The assumptions that individual decision-makers are the ultimate judges of their own well-being and that their "tastes" (the fulfillment of which determines their well-being) are exogenous to the economic system reflect deep philosophical convictions. Both of these assumptions are rooted in the Enlightenment view of man as free, autonomous, and possessed of the unalienable rights to life, liberty, and the pursuit of happiness. There is no doubt that this intellectual stance (which also underlay 19th-century liberalism and its 20th-century intellectual descendants) has a number of positive political consequences. Policy makers informed by this view will be very hesitant to override individuals' preferences as manifested in their market choices. The placing of consumers' tastes off limits to state manipulation can serve as a barrier to overly ambitious (and intrusive) schemes for social engineering.

Regardless of the political advantages of this philosophical orientation, the problems with the rational actor paradigm as an *analytical* starting point remain. In order to retain optimization as a basic modelling assumption, neoclassical economists often claim that people behave "as if" it were true. There is a long-standing methodological debate within economics about whether it is appropriate to model firms and consumers "as if" they were capable of perfect maximization and had full information about the market and technological possibilities available to them. Actual practice suggests that there has been a widespread adoption of the "as if" premise as a starting point for modelling, so the basic question of the legitimacy of this stance is rarely raised explicitly. Yet it remains a nagging source of doubt about the validity of theoretical or predictive economic models. In cases such as climate change, where the results of the analysis really matter for practical policy design, it is worth revisiting this foundational question.

A canonical statement of the "as if" approach to economic reasoning is given by Milton Friedman in his influential essay, "The Methodology of Positive Economics" (1953). It is worth quoting Friedman at length, because his approach is so much a part of the unexamined practice of modern-day economic modelling:

> [U]nder a wide range of circumstances individual firms behave *as if* they were seeking rationally to maximize their expected returns (generally if misleadingly called "profits") and had full knowledge of the data needed to succeed in this attempt; *as if*, that is, they knew the relevant cost and demand functions, calculated marginal cost and marginal revenue from all actions open to them, and pushed each line of action to the point at which the relevant marginal cost and marginal revenue were equal. Now, of course, businessmen do not actually and literally solve the system of simultaneous equations in terms of which the mathematical economist finds it convenient to express this hypothesis....[T]he businessman may well say that he prices at average cost, with of course some minor deviations when the market makes it necessary (pp. 21-22, footnote omitted, emphasis in the original).

It is interesting that Friedman goes on to justify this position by making a sophisticated appeal to what amounts to an *evolutionary* argument. He says:

> Confidence in the maximization-of-returns hypothesis is justified by evidence of a very different character [that is, different from the testimony of businessmen]....[U]nless the behaviour of businessmen in some way or other approximated behaviour consistent with the maximization of returns, it seems unlikely that they would remain in business for long. Let the apparent immediate determinant of business behaviour be anything at all – habitual reaction, random chance, or whatnot. Whenever this determinant happens to lead to behaviour consistent with rational and informed maximization of returns, the business will prosper and acquire resources with which to expand; whenever it does not, the business will tend to lose resources and can be kept in existence only by the addition of resources from outside. The process of "natural selection" thus helps to validate the hypothesis – or rather, given natural selection, acceptance of the hypothesis can be based largely on the judgment that it summarizes appropriately the conditions for survival (ibid., p. 22).

What is remarkable about the second part of Friedman's argument is what is left unsaid. While it is undeniably true that "natural selection" in a competitive marketplace will tend to reward more profitable firms and punish less profitable ones[28], there is nothing in evolutionary theory that guarantees that the population of firms *at any particular point in time* will behave as if optimization on all margins had been achieved. Evolution and natural selection occur in real time, which means that even if some hypothetical optimum would *eventually* be achieved by evolving firms, it need not be the case that the firms are optimized *now*. Indeed, the notion that evolution guarantees optimization at all times is internally inconsistent – evolution is continuous, and if optimization were assured, then the configuration of firms *at all*

28 "More profitable" does not necessarily imply "more rational." Thaler (2000) points out that the evolutionary argument which economists often resort to is false; he discusses a model by De Lang et al. (2000) in which less rational traders in a financial market carry off a greater reward than the more rational ones.

past times would also have been optimal, so that no evolutionary progress would be required. On the other hand, if market, technological, or regulatory conditions are changing over time, then the question of how rapidly evolutionary change brings about adaptation to the changing conditions is an empirical one, and cannot be answered without consideration of the actual rates of change that are possible, the distribution of "genetic information" in the population that provides the material for emergence of new forms, etc.

Furthermore, evolution, whether biological or economic, is a path-dependent process. That is, the incremental changes that emerge under the pressures of natural selection build on structures that have previously evolved, and hence reflect the past adaptation to prior conditions that may no longer persist. Path-dependent evolutionary processes are not in general expected to produce "optimal" outcomes. As Nelson has said, "[a]ny 'optimizing' characteristics of what exists...must be understood as local and myopic, associated with the particular equilibrium that happens to obtain" (1995, p. 51). In both nature and the economy, we find a considerable amount of variation in a given species' (industry's or sector's) population. Variation in a population itself may be an evolutionary adaptation – variation can create resilience to environmental changes (or its analog in a market setting, the rate at which organizations can change their practices in response to changing market conditions). Yet variation itself is almost certain to produce differences in "efficiency" if efficiency is measured along a single dimension of performance. This has recently been demonstrated in models that exhibit diversity in the organizational form of firms, diversity that, among other things, can account for phenomena such as the "Porter hypothesis" that environmental regulation of the right kind need not adversely affect the economic performance of the regulated firms (Porter, 1991; Porter and van der Linde, 1995; DeCanio et al., 2000b).

Thus, we see that the "as if" argument must necessarily rest on some additional theory that actually explains how optimization might emerge through selection. The correspondence between the "optimization" that results from the evolutionary process and optimization-as-conscious-calculation then depends on the efficacy of the underlying evolutionary mechanism. It is probably too much to require, at the present stage of our knowledge, that any truly scientific theory be able to spell out the underlying process with the precision of an algorithmic computation, although it can be argued that such an ideal should be the ultimate objective of theory (and will carry its own set of limitations and constraints on what can be known and what can be done) (Rust, 1997; DeCanio, 1999). Yet the quest for better and better representations is an important part of the scientific enterprise. As Herbert Simon put it in a symposium responding to Friedman's approach to economic methodology, the task of economics should be to "make the observations necessary to discover and test true propositions.... Then let us construct a new market theory on these firmer foundations" (1963, p. 230).

Acknowledging that rationality is bounded is an important step in such a process. Recent research on decision-making algorithms employed by firms demonstrates that as theory moves in the direction of greater realism, the reasons for believing in bounded rationality become stronger. In particular, paying attention to the organizational structure of firms in even the simplest kind of models leads to characterizations of firms' decision algorithms that are boundedly rational. The myriad of potential organizational forms precludes direct calculation of the optimal one, and evolutionary processes or plausible search heuristics that produce improvements in performance are unlikely to find anything other than local maxima (DeCanio et al., 2000a, 2000b, 2000c).

For our present purposes, the crucial question is whether the behaviour that is actually carried out by the economic agents has different consequences for economic modelling of climate policy than the "as if" presumption of maximization. The tentative answer to this question appears to be "yes". This translates into recognizing the limitations on existing models that attempt to describe the "macroeconomic" or economy-wide costs of carbon-reduction policies. Section 2.2.5 provides an illustration.

2.2.5. *Relaxing the assumption of the optimizing firm: Implications*

If we relax just one assumption of standard neoclassical models, allowing for bounded rationality on the part of firms, and otherwise stay entirely within the neoclassical framework, we can easily illustrate how existing models that assume an efficient utilization of resources have misleading policy conclusions.

Figure 1 represents schematically the relationship between the GDP of an economy and its CO_2 emissions[29]. Economic activity results in both the production of "goods" (represented here by GDP) and "bads" (pollution in general, represented here by CO_2 emissions). The family of curves U_iU_i are the "social indifference curves" that reflect preferences over both the explicit (market) value of the goods and services produced by the economy and the implicit disutility of CO_2 emissions[30]. For a given level of CO_2 emissions, higher GDP is associated with higher social utility, while for a given level of GDP, higher utility is associated with *lower* CO_2 emissions. Each U_iU_i curve represents the combinations of GDP and CO_2 emissions yielding the same level of social utility, and utility increases with the index *i* of the U_iU_i curve. These curves are upward sloping because the disutility of higher CO_2 levels can only be compensated for by a higher level of ordinary GDP.

The curve AA represents the production-possibilities frontier of the economy. This curve is borrowed straight from neoclassical economics. It describes the trade-offs between GDP and CO_2 reductions that would result if firms were optimizing and if production technologies were convex[31]. Neoclassical models assume that all the

---

[29] Figure 1 follows the treatment given in Sanstad et al. (2000a).

[30] Allow for purposes of this argument that such an indifference map could be constructed. Economists since the 1930s have been reluctant to make welfare judgments founded on interpersonal comparisons of utilities (Slesnick, 1998). The fact that societies do make "choices" is not equivalent to the existence of a social indifference map with properties analogous to those of an individual. This will be discussed further below.

[31] If one allowed for non-convexities in production, this curve could look different, for example, showing wiggles. That is, even if firms were fully optimized, there could be instances where a shift along the curve would increase GDP *and* reduce CO_2 emissions.

producers in the economy are fully optimized and technically efficient, i.e., on the curve. Given the shape of the underlying production technologies, a reduction in CO_2 can only be achieved at the cost of giving up GDP.

In reality, however, producers are not completely efficient. As noted earlier, large numbers of empirical studies of production efficiency have established that deviations from full technical efficiency are common throughout the economy.

With regard to energy efficiency in particular, the literature on the "energy-efficiency gap" demonstrates that complete productive efficiency is not being reached in the provision of energy services either. (See, for example, Interlaboratory Working Group, 1997, 2000; Office of Technology Assessment, 1991; National Academy of Sciences, 1992; Tellus Institute, 1997; Krause et al., 1999; and Bernow et al., 1999).

The failure of firms (and hence, the economy as a whole) to achieve complete technical efficiency can be attributed to, among other things, the pervasiveness of bounded rationality. Regardless of the particular cause or causes of the limitations, bounded rationality implies that the economy is initially (now) at point I, interior to the production-possibilities frontier. Policy measures, organizational improvements, or gains in productivity from assimilating better practices found in other firms can result in movements away from the initial point I, potentially all the way to the actual production-possibilities frontier. In Figure 1, four such movements are shown, to points M_1, M_2, M_3, and M_4.

The movement from I to M_1 is feasible (as are all the movements to the points M_i) because the economy remains within its production-possibility frontier. In moving to M_1, the economy increases its output of ordinary goods and services while also increasing its CO_2 emissions. However, this "improvement" in productive efficiency is accompanied by a drop in social utility, because M_1 is on a social indifference curve (not drawn) that lies to the right of the social indifference curve U_0U_0 passing through I.

The movement from I to M_2 represents a change in which GDP is increased and CO_2 emissions are decreased. It also constitutes an improvement in social utility, because the indifference curve U_1U_1 is to the left of U_0U_0. Because the utility level of M_2 is higher than that of point I, the movement to M_2 corresponds to a "Pareto-improvement" for the economy, that is, it is a movement such that at least some members of the economy are better off at M_2 than at I, and no one is disadvantaged by the move. It is still not a movement all the way to the production-possibilities frontier, however.

The potential moves to M_3 and M_4 represent efficiency improvements that do reach the frontier of technical production efficiency for the economy. Once the frontier is attained, there is an "opportunity cost" to increasing GDP or reducing CO_2 emissions any further. That is, on the frontier an improvement in either dimension of economic performance can only be achieved at the expense of performance in the

Figure 1: The Production Possibilities Frontier and the Relationship between GDP Change and CO_2 Reductions

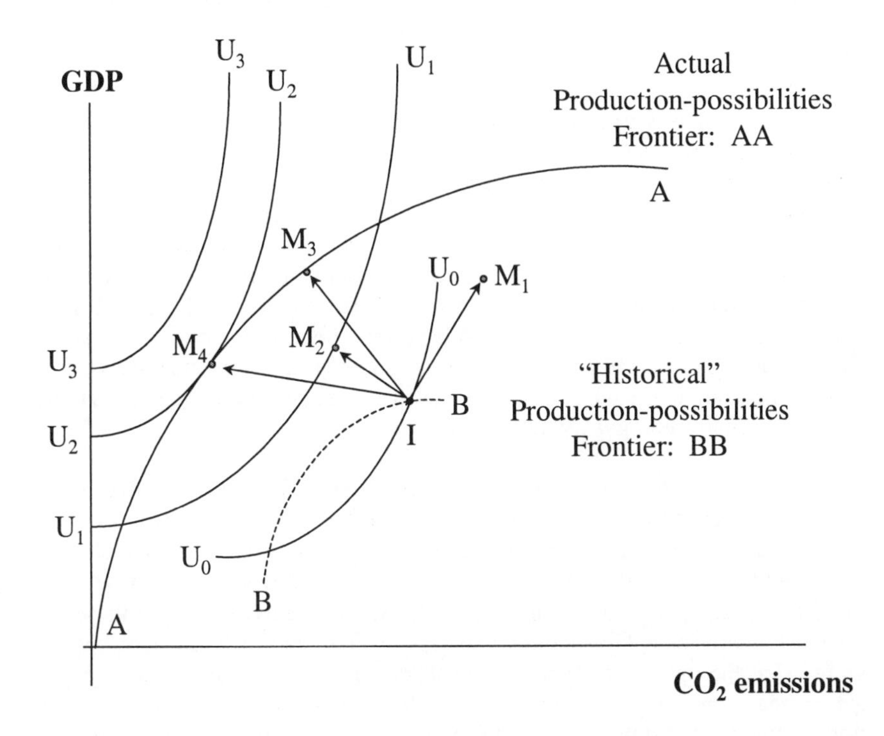

other dimension. It should be noted, however, that even given an opportunity cost in moving from M_3 to M_4, (the reduction in CO_2 emissions entailed by such a move can be obtained only at the expense of some GDP), social utility is higher at M_4 than it is at M_3. The point M_4 indicates the maximum attainable utility (given the productive capacity of the economy); the utility curve U_2U_2 passing through M_4 is tangent to the production-possibilities curve AA. (It is this tangency of the utility curve to the production-possibilities frontier that characterizes the highest attainable utility level given the technological possibilities.)

One final feature of Figure 1 needs to be explained. The dotted curve BB drawn through point I is what might be designated the "historical" production-possibilities frontier. It is the frontier that might be found by econometric techniques utilizing data on the economy's past performance. The past performance of the economy would reflect the existing efficiency gaps, that is, past performance would embody the instances of bounded rationality, organizational failure, etc., that presently exist in the economy. It would be possible to interpret the historical frontier BB as measuring the slope of a tradeoff between CO_2 emissions and GDP, but such an interpretation would be misleading. Historical data might suggest that no reductions in emissions could be achieved without a cost in terms of ordinary output, but those data do not take account of all the possible efficiency-improving moves that the economy actually is capable of. The failure of most existing integrated assessment models to incorporate such possibilities is a source of bias in their estimates of the cost of carbon reductions. The GDP loss that would be required to achieve any given cut in CO_2 emissions by moving along curve BB is not indicative of what could be achieved by policies that would move the economy towards points like M_2, M_3, or M_4[32].

At present, there is neither a carbon tax nor mandatory carbon emissions permits (in the United States), and thus no cost to emissions of CO_2 from combustion of fossil fuels. As Figure 1 is drawn, the point I is at the maximum of the "historical" production-possibilities curve BB, so that the slope of BB is zero at I. The tangent to BB at I is horizontal, corresponding to a zero price for CO_2 emissions at I. Imposition of a carbon tax (or implementation of a tradable emissions permit system) would have the effect of making this relative price line positively sloped. Introducing a price for carbon emissions could have the effect of inducing movement away from point I towards the actual frontier, by stimulating high-return R&D, focusing management attention on correctable inefficiencies, or increasing the salience of already-profitable energy-saving technologies.

Whether or not a price-only policy achieves the socially optimal amount of carbon emission reductions depends on two things: (1) the price would have to be set optimally, equal to the slope of the production-possibility frontier and the slope of the social indifference curve at M_4, and (2) the price signal would have to be sufficient, in itself, to push the economy all the way to AA. In reality, neither of these requirements is guaranteed. Given the present state of scientific knowledge, we do not know what the optimal carbon emissions price is (we can only be certain that it is greater than zero) and we do not know all the factors – social, behavioural, informational – that could bring about Pareto-improving moves towards the frontier.

The discussion in this section falls entirely within the premises of the neoclassical paradigm, except that Figure 1 allows the economy to be operating inside its production-possibilities frontier. The existence of bounded rationality (and the resulting inefficiency) in production is not the only way that neoclassical models can lead to biased and misleading results in climate policy analysis, however. Corresponding to the failures of maximization that have been documented on the production side of the economy, there are numerous reasons to believe that consumers fail to optimize when it comes to the choice of technologies that have differential impacts on atmospheric carbon loadings.

2.2.6. *The individualist stance and co-evolution of tastes and culture*

As useful as they may be for some kinds of analysis, the assumption of atomistic consumers with immutable tastes is a fiction. As we will elaborate later (in section 2.2.8), consumers' utilities are as likely to be interdependent as independent. Even

---

[32] Compilations of the kinds of policies that could accomplish such Pareto improvements include Interlaboratory Working Group (2000, forthcoming), Krause et al. (1999), Energy Innovations (1997), Geller et al. (1999), and Hoerner and Gilbert (1999).

more fundamentally, the tastes are, at least to some degree, the result of a complicated process of education, socialization, and adaptation that is dependent on (among other things) the direction taken by technological change and the evolution of the material culture.

Even if human needs for food, shelter, mobility, security, and self-esteem are immutable, the way in which these needs are met by tangible commodities and particular services can vary considerably. Consider mobility. In the modern context, this means the ability to travel considerable distances cheaply, conveniently, safely, and in comfort. Obviously, there are different kinds of vehicles that can provide services of this type – automobiles, bicycles, buses, trains, aircraft, and ships or boats. The attributes of these different kinds of vehicles vary across categories and within categories; within a particular category, different vehicles can offer a continuum of features along each of the desirable dimensions, depending on design features.

This can be illustrated by examination of the safety attribute. Safe transportation can be achieved by increasing the crashworthiness of vehicles or reducing the likelihood of crashes. The probability of crashes can be reduced by employment of specialist operators (e.g., airline pilots) or by improving the controllability of the vehicles themselves (disc brakes, modern radial tires, collision warning systems, etc.) Design changes and/or behavioural incentives can improve crashworthiness. Thus, while the introduction of safety belts in automobiles was an important safety innovation, enforcement of fines for failure to wear the belts has proven to be a very low-cost way to increase safety belt use (in the United States, at least). At the same time, automobile safety depends on complex interactions of behavioural and technological factors. The correlation between crashworthiness and vehicle weight is partially offset by the greater recklessness of drivers of high-weight vehicles (who at some level recognize the advantage of their vehicles in collisions with lighter cars). The presence of lightweight and heavyweight vehicles on the road produces an externality. Passengers in high-weight vehicles have a much better chance of survival in collisions with low-weight vehicles than the occupants of the low-weight vehicles (Evans, 1985; Bradsher, 1997). This constitutes an involuntary transfer of risk from the occupants of high-weight vehicles onto the occupants of low-weight vehicles. In turn, the externality leads to an escalation of vehicle weights as car buyers seek to protect themselves and their families. Thus, the same level of safety achieved by the existing automobile fleet could be achieved under a weight limitation policy (or alternatively, higher fuel economy standards) and the redirection of automobile design resources into other safety features such as passenger protection, impact energy absorption, and highway barrier "softening".

This example illustrates how a consumer preference for safety need not take the particular form it does – as in the case of the proliferation of truck-based sport utility vehicles (SUVs) in the United States[33]. The particular form in which this consumer preference has become manifest is dependent on a large set of path-dependent tech-

---

[33] Of course, some consumers desire other attributes of SUVs such as carrying capacity and visibility. However, different engineering routes could also reach these objectives.

nological developments and policy choices. These include – but are not limited to – decisions to subsidize road building, laws regulating automobile insurance, choices of which safety attributes to regulate (e.g., mandating performance in single-vehicle crash tests but not prohibiting "aggressiveness" towards vehicles that might be struck in multi-car collisions), and so forth. To assert that the mix of products being produced at any given time reflects the "optimal" way of meeting consumers' needs begs the question of what path society has taken to arrive at the particular solution to set of attribute demands that could have been met in many other ways[34].

This argument is similar to the discussion (in Section 2.3.1 below) about the distinction between energy *consumption* and energy *services* in meeting consumers' and producers' demands (Howarth, 1997). Models in which energy consumption *per se* is the source of utility or productive potential will show different responses to greenhouse-gas limiting policies than models in which it is the provision of energy services that matters. If fuels are a direct input to the production functions of firms, then any policy that raises the price of those fuels will automatically have an adverse effect on the economic performance (profitability) of the firms. On the other hand, if it is energy services that are demanded, and those services can be provided in more or less energy-efficient ways, then the flexibility of firms in responding to policies to change the rate of CO_2 emissions will be greater than in a model that specifies direct fuel inputs (Hanson and Laitner, 1999).

A given set of preferences can be satisfied in different ways, and it is also true that the tastes themselves are *social phenomena*, not "laws", human or natural. To continue with the automobile example, the demand for mobility may be deeply rooted in the human psyche, but the attitudes and connotations surrounding the automobile are to a large degree socially constructed. Cars serve many psychological and relational functions in addition to being a mode of transportation – they display wealth and success, embody feelings of power and control, constitute a form of courtship display, and are an affordable type of high design art. The agglomeration of these functions in the automobile is not determined by any inexorable social dynamic. Other technologies and commodities have offered some or all of these functionalities in the past (horses, hunting falcons, expensive suits of armor, etc.). It is possible to imagine other goods (some of which have a smaller ecological footprint) serving some or all of the same functions in the future. High-performance personal computers could be imagined to convey a great deal of the same kinds of symbolism, for example.

2.2.7. *Microfoundations, aggregation, and the uniqueness and stability of equilibrium*

General Equilibrium theorists have not been oblivious to potential obstacles in applying the theory to policy analysis, and in this section we address several such issues. For example, the problem of aggregation – moving from a description of the economic behaviour of individuals to a description of the characteristics of the

---

34 Buying an SUV may not even be an effective way of purchasing safety. According to one news report, "[s]port utilities are three times as likely as cars to kill the other driver in a crash, but the death rate for sport utility occupants is just as high as for car occupants because of sport utilities' tendency to roll over and their lack of crumple zones" (Bradsher, 2000, p. C2).

macroeconomy made up of those individuals – casts severe doubt on the typical formulation of demand and supply functions in applied CGE models. We briefly describe these difficulties and offer implications for policy applications.

To understand the motivation for formulating CGE models as they are usually set up nowadays, let us return to the "microfoundations" theme. As already mentioned, there has been, over the last couple of decades, an impetus in academic economics to place the entire theoretical edifice of the discipline on "microfoundations" – that is, to make the theory of all branches of economics, especially macroeconomics, analytically consistent with the model of the individual economic agent[35].

The search for a macrotheory that derives the behaviour of aggregates from the behaviour of their constituent micro-units is an attractive, yet daunting and ambitious, project. Nor is it confined to economics alone. In the natural sciences, this project is far from completed, if completed it can ever be. For example, models of the behaviour of large molecules are not necessarily consistent with, or derived from, models of the behaviour of elementary particles. Different theories of matter are formulated for different levels of aggregation of matter (let alone living things)[36].

In economics, the dominant theory of micro-behaviour is the neoclassical theory of consumer and producer choice. Thus, the effort to provide microfoundations for macroeconomics attempts to explain macro-phenomena in terms of this microeconomic model of consumer and producer behaviour. In most CGE models, this is done by simply *assuming* that macro-behaviour displays the same features as micro-behaviour[37]. For example, it is assumed that market demand, which conceptually is the sum of individual consumer demands, has the same mathematical properties as the demand function of an individual consumer maximizing utility under a budget constraint. This so-called "representative agent" device is used on the consumption side (through the "representative consumer") as well as on the production side (through the "representative producer" in each production sector)[38].

---

35 See Weintraub (1979), Janssen (1993), and Rizvi (1994). The latter two offer an historical perspective on the microfoundations project.

36 The popularity of the purely reductionist worldview that this should be possible, at least in principle, is waning. Deutsch (1997, p. 175 ff.) provides a window on the current discussion of this topic in physics.

37 Some readers may object to the term "macro" in conjunction with CGE models, as they differentiate between sectors of the economy and allow the exploration of alternative resource allocations. This topic is traditionally considered the domain of microeconomics, while the term "macroeconomics" is reserved for the study of national accounting aggregates like production, consumption, income, investment, the trade balance, and employment, as well as the consideration of monetary phenomena and policies. We could speak of CGE models as *multisectoral models*. The point here is that functions in CGE models represent *groups* of economic agents, not individuals.

38 Several CGE models distinguish between groups of consumers – more accurately: households – that differ by income or another socioeconomic characteristic. In this case, there is one representative consumer for each such group.

In the 1970s, some general and far-reaching theoretical results were developed showing that this approach lacks internal consistency. Sonnenschein (1972, 1973), Mantel (1974) and Debreu (1974) proved that it is only in highly restrictive cases that aggregate demand – the sum of individual, utility maximizing consumers' choices – has the same mathematical properties as individual demand[39]. Consumers have to be virtually identical in their preferences and endowments for this to happen. As soon as one allows for some heterogeneity in consumer preferences or endowments, aggregate demand can take on almost any shape. Shafer and Sonnenschein, in their influential survey of these results (1982), conclude:

> [S]trong restrictions are needed in order to justify the hypothesis that a market demand function has the characteristics of a consumer demand function. Only in special cases can an economy be expected to act as an 'idealized consumer' (p. 672).

To illustrate the aggregation problem, Kirman (1992, p. 124 f.) provides an example created by Jerison (1984): Two individual agents are each equipped with a standard neoclassical formulation of preferences. Faced with two different budget constraints X and Y, they both prefer their utility-maximizing consumption bundle chosen under X to the one under Y. The "representative" individual that is constructed as the sum of their choices, however, prefers the consumption bundle that would be chosen under aggregate budget constraint Y to the one under aggregate budget constraint X. Thus, going from the individual agents to the "representative" agent can imply a reversal of preference order! But this is a somewhat special example. The more important problem is that aggregate excess demand can display properties that contradict those of individual demand functions, as they are stipulated and relied on by microeconomic theory. We return to this point later[40].

These results render the device of the "representative agent" in applied general equilibrium analysis highly questionable. They imply that even if one accepts the utility-maximizing consumer as a model for *individual* decision making, it is not valid for *aggregate* decision making. Since CGE models deal with aggregates (they describe market demand and supply functions), their use of micro-functions is inconsistent. Put differently, it is only consistent if one makes the explicit assumption that consumers and producers are virtually identical. But this is clearly at odds with reality.

---

39 To be exact, this was first shown for *aggregate excess demand* (the difference between aggregate supply and aggregate demand), then for *aggregate demand* (see Shafer and Sonnenschein, 1982). As we will later explain, it is the aggregate *excess* demand function that creates technical problems for model solutions and theory. The aggregation problem applies to both.

40 There is also an aggregation problem on the production side. Franklin Fisher has shown that for an aggregate production function (i.e. the sum of individual production functions) to exhibit neoclassical properties, the individual functions have to satisfy very stringent conditions (1968a, 1968b, 1969). He concludes: "Aggregate production functions practically never exist" (1982, p. 136).

Stoker (1993) characterizes the situation as follows:

> Taken at face value, representative agent models have the same value as traditional, ad hoc macroeconomic equations; namely they provide only statistical descriptions of aggregate data patterns, albeit descriptions that are straightjacketed by the capricious enforcement of restrictions of optimizing behaviour by a single individual (p. 1829).

Over the years, several economists have expressed unease about what they observed to be a veritable pressure to employ neoclassical "microfoundations". Work not using the theoretical apparatus of utility and profit maximization came to be characterized as "ad hoc". In 1986, James Tobin complained (quoted in Rizvi, 1994)

> [The microfoundations] counter-revolution has swept the profession until now it is scarcely an exaggeration to say that no paper that does not employ the 'microfoundations' methodology can get published in a major professional journal, that no research proposal that is suspect of violating its precept can survive peer review, that no newly minted Ph.D. who can't show that his hypothesized behavioural relations are properly derived can get a good academic job (p. 350).

Martel (1996) observes poignantly that Keynesians, who were true macroeconomists in the sense of "developing a macro-behavioural theory of the relationship between aggregates, guided by any significant historical patterns in the aggregate data ...[,]" something "considered a respectable thing for an economist to be doing until the 1970s[,] ... began to be indicted for *ad hocery* in neoclassical courtrooms" (emphasis in the original, p. 131).

Why does the economics profession cling to the representative agent device? Giving it up would open the door to technical problems with applied models, as well as problems with some central tenets of General Equilibrium Theory, if not today's economics itself. We take a little space to describe these problems, though our rendering of these issues will necessarily be oversimplified and incomplete.

The aggregation problem inherent in the Sonnenschein-Mantel-Debreu results strikes at the very heart of General Equilibrium Theory[41]. The essence of that theory can be summarized by the condition that aggregate excess demand be zero. Aggregate excess demand is a vector whose elements represent the difference between demand and supply in each market. That this vector be strictly zero – i.e., that each of its elements be zero – is a formal statement of the condition that all markets clear in equilibrium[42]. Market clearance is brought about through prices that move so as

41 This passage draws on the excellent and insightful exposition of Ingrao and Israel (1990). Ackerman (1999) provides an acute and provocative treatment of the issues discussed here.

42 If one allows for free goods, excess demand for those goods may be negative, i.e., supply may exceed demand.

to eliminate excess demand in each market. Thus, aggregate excess demand is a function of prices. A price vector that makes this function zero is an equilibrium price vector.

The shape and the properties of this excess demand function are of central importance for General Equilibrium Theory. In particular, one is concerned about whether the function has a zero solution at all (this would guarantee the *existence* of equilibrium), whether it has just one or many zero solutions, and if the latter, whether they lie on continuous stretches or are discrete (this corresponds to *uniqueness* of equilibrium and its stability). Existence of equilibrium clearly is a necessary condition for the theory to make sense. Uniqueness is desirable because without it, the comparative statics analysis that is the content of most policy evaluation – comparing the benchmark equilibrium of the reference case to the "counterfactual" equilibrium resulting from a policy shock – would become inconclusive. If a model economy with multiple equilibria were perturbed by a policy intervention and could move to any one of several equilibria, one could not draw an unambiguous conclusion about the effect of the policy.

It turns out that the conditions for existence of general equilibrium are fairly general, but the situation looks very grim for uniqueness. Theoretical economists have undertaken great efforts to examine and successively weaken the assumptions about the shape of the aggregate excess demand function that are needed to ensure uniqueness of equilibrium, but with little success. For a long time now, the profession had to acknowledge that the presence of multiple equilibria cannot be ruled out; in fact, there are indications that multiple equilibria are very common[43]. As to applied modelling work, Kehoe (1998) observes "It may be the case that most applied models have unique equilibria. Unfortunately, however, these models seldom satisfy analytical conditions that are known to guarantee uniqueness, and are often too large and complex to allow exhaustive searches to numerically verify uniqueness" (p. 39).

The effort then turned to examining the conditions for *local* uniqueness – for the equilibria to be discrete, i.e., not to lie on a continuum, and for their number to be finite. Local uniqueness would be a minimum requirement for the plausibility of comparative static exercises that are confined to examining small changes to the status quo (provided that model economies which are close to each other, i.e., which have similar initial endowments and parameterizations of algebraic functions, have sets of equilibria that are also close to each other)[44]. But even this more modest program – seeking to ascertain the legitimacy of assuming "regular" economies with multiple, but finite and well-defined equilibria – is not guaranteed success. It has been shown that "even in economies where there are usually a finite number of equilibria – such as economies with a finite number of consumers – the addition of distortionary taxes or externalities can lead to robust indeterminacy" (that is, con-

43 Kehoe (1998) provides a simple example of a two-consumer, two-commodity pure exchange economy, whose consumers have standard neoclassical utility functions, which has three equilibria (p. 39 ff., p. 55 ff.).

44 This is the concept of "regular economies" which was created by Gérard Debreu (1970).

tinua of equilibria that are not eliminated by small perturbations to the model economy)[45].

It is one thing to assess the number of equilibria, another to examine the processes by which prices move the economy towards these equilibria (or fail to do so). This issue pertains to the *stability* of equilibrium. How prices would adjust spontaneously is an important question, as surely few economists alive today – and few in the body politic at large – would argue for equilibrium prices to be centrally imposed[46]. Rather, the majority can safely be assumed to have faith that the market, and the free play of the price system, is an effective mechanism to generate, if not efficient resource allocations in the sense of full Pareto-efficiency, then at least allocations that are superior to those resulting from central planning. Ingrao and Israel (1990) put it this way:

> An ideological standpoint that regards the market as possessing the virtue or intrinsic property of combining subjective behaviour harmoniously cannot content itself with simply knowing that a final state of equilibrium exists. It has to show that the economy is capable of attaining this state spontaneously, that the system's variables of state – i.e., prices – vary and adjust in such a way as to arrive at a vector of equilibrium prices. Otherwise, one would be forced to acknowledge that market forces are not capable of leading the market itself to equilibrium and that Smith's "invisible hand" wavers Sisyphus-like around the actually existing equilibrium position without having the strength to push the economic system into it (p. 331).

Theoretical economics has not ignored the process of price adjustment. Léon Walras, the father of General Equilibrium Theory, suggested the tâtonnement process that we mentioned earlier (Section 2.1). While the hypothetical auctioneer, who adjusts and determines prices before trading ensues, is a far stretch of the imagination, the principle according to which he would adjust prices is intuitively appealing: Positive excess demand for a commodity would increase its price, negative excess demand would lower it[47]. However, during the 1960s and 1970s, the tâtonnement price adjustment process was shown to lack convergence in a general equilibrium setting[48]. The Sonnenschein (1972) result that the shape of aggregate excess demand is arbitrary had implications not only for uniqueness of equilibrium, but also for the stability of the price adjustment process, because an important condition for the stability of the Walrasian price adjustment process – the gross substitutability con-

---

45 Kehoe, Levine and Romer (1992), quoted in Kehoe (1998), p. 83.

46 Ingrao and Israel (1990) comment on writers who have been sympathetic to this idea (p. 333).

47 Samuelson (1941) was the first to write down this process in formal terms.

48 We follow the account given in Ingrao and Israel *ibid.*, Chapter 12.

dition[49] – does not follow from the shape of the individual demand functions. Thus, there is no reason to assume that this condition should hold.

Economists began to explore other price adjustment mechanisms, for example those that allow trading out of equilibrium. But this can change the initial conditions that are used to compute equilibrium. Equilibrium becomes dependent on the path taken to reach it (see, for example, Fisher 1983). There are many other approaches trying to develop price adjustment mechanisms that lead to equilibrium, but they rely on unrealistic and elaborate assumptions about consumer properties and institutional arrangements.

Some rather sweeping results on this topic are due to Donald Saari and his collaborators[50]. They showed that price adjustment mechanisms that rely only on information from the aggregate excess demand function do not generally converge. Saari and Williams (1986) summarize:

> [U]nless unrealistic informational requirements are imposed upon the system, the standard economic assumptions need not lead to convergent, decentralized dynamics. The instability of these economic systems must either be accepted, or else the hypotheses that lie behind them must be reexamined (p. 153, quoted in Ingrao and Israel 1990, p. 443).

Franklin Fisher, one of the leading theorists in the field, has summed up the state of affairs in a survey on the stability problem as follows:

> [W]e have no rigorous basis for believing that equilibria can be achieved or maintained if disturbed....[T]here is no disguising the fact that this is a major lacuna in economic analysis....[N]ew modes of analysis are needed if equilibrium economic theory is to have a satisfactory foundation (1989, pp. 36, 42).

More recent work seeks to identify price adjustment mechanisms that rely on more information, i.e. the value which prices take on in the course of the adjustment process[51]. But these studies still require elaborate manipulations on the part of the hypothetical auctioneer. No adjustment process has yet been modeled that would follow from plausible behaviour of the individual agents that make up the market[52].
Clearly, this is a disturbing result for macroeconomic models with neoclassical foundations. If the equilibrium is not stable, then focusing attention on equilibrium

49 See Arrow, Block and Hurwicz (1959).

50 Saari and Simon, 1978; Saari, 1985; Saari and Williams, 1986.

51 Van der Laan and Talman (1987), Herings (1996 and 1997).

52 However, interesting empirical work is being generated that could help explain real-world price adjustment processes. For example, Blinder et al. (1998) explore price setting behaviour by firms and find that in many cases, it is governed by considerations quite different from the ones stipulated in neoclassical theory.

analysis may be misplaced[53]. Of course there are macroeconomic models that attempt to capture some "disequilibrium" phenomena, like involuntary employment. These are to be welcomed, in the sense that they offer alternative perspectives, especially if they do not pretend to excessive reliance on microeconomic theory. However, these models, too, cannot offer an entirely satisfactory description of how prices adjust. Also, their role in climate change mitigation assessment has diminished as Computable General Equilibrium models have become more prevalent[54].

There have been a number of efforts to rescue general equilibrium theory[55]. A few authors have tried to salvage general equilibrium analysis by tackling the aggregation problem head on. For example, Werner Hildenbrand has explored the possibility that the *distribution* of individual household characteristics could generate properties of aggregate excess demand that would ensure uniqueness and stability of equilibrium. Looking at aggregate expenditure data from France and the U.K., he finds considerable empirical support for his hypotheses (1994). This is no doubt an interesting approach. Exploring the distribution of individual agents' characteristics is a first step to considering phenomena that are "macro", i.e. phenomena that do not exist at the individual level. Indeed, Hildenbrand addresses the importance of focusing on macro-phenomena (which most economic data reflect), as distinct from micro-theory:

> I believe that the relevant question is not to ask which properties of the individual demand behaviour are *preserved by* going from individual to market demand, but rather to analyze which new properties are *created* by the aggregation procedure (1994, p. ix. f., emphasis in the original).

But Hildenbrand's micro-units are still atomistic individuals who do not interact directly and are not influenced by the behaviour of macro-aggregates. Accounting for interaction among the agents would move further in the direction of finding true microfoundations for macroeconomics. Yet another step could be allowing for macro-level variables to exert an influence on micro-behaviour (in the words of David Colander (1996b), providing "macrofoundations for micro" – that is, microeconomics). Synergetic opinion formation models, employed in sociology already do this in some form (Weidlich and Haag, 1983). Kirman (1989) observes:

---

53 Our discussion does not by any means exhaust the literature on dynamics and instability. For example, models in which *expectations* are important (that is, virtually any model in which forward-looking planning is part of the agents' decision-making process) suffer from instability for reasons distinct from those discussed in the text. According to Grandmont, "[l]earning, when agents are somewhat uncertain about the dynamics of the economic or social system, is bound to generate local instability of self-fulfilling expectations, if the influence of expectations on the dynamics is significant" (1998, p. 742).

54 See also Ginsburgh and Keyzer (1997) for references to CGE models with disequilibrium in some markets.

55 Ackerman (1999) offers a lucid account and interpretation of these efforts.

> [I]t is clear that in the standard framework we have too much freedom in constructing individuals. The basic artifact employed is to find individuals each of whose *demand behaviour* is completely independent of the others. This independence of individuals' behaviour plays an essential role in the construction of economies generating arbitrary excess demand functions. As soon as it is removed the class of functions that can be generated is limited. Thus making *individual behaviour* dependent or similar may open the way to obtaining meaningful restrictions (p. 138 f., emphasis in the original).

It is heartening to see that research in this spirit is beginning to be produced. (See, for example, the programmatic article "The economics of heterogeneity", by Delli Gatti, Gallegati, and Kirman (2000), and the work presented in the associated volume (2000).)

The reason this theoretical discussion matters for appraisal of climate policy analysis is that the CGE models currently in use for that purpose simply ignore these problems. The representative agent framework is standard, and the question of what the real economy's path of adjustment to a change in the policy environment might be is not considered. The methods used to compute the "general equilibrium" are purely mathematical algorithms that do not reflect any kind of underlying behaviour by the individuals and organizations making up the economy. Hence, their representation of economic reality is at best only partial; at worst it is entirely inaccurate and unreliable.

2.2.8. *Welfare measurement, distribution, and intergenerational equity*

Economics is often looked upon as the ultimate arbiter of policy choices, because it seems to offer something the other social sciences do not: a framework capable of valuing the consequences of different policy choices with a single metric. The most common approach takes GDP or one of its variants as a measure for aggregate well-being. This is clearly inadequate, as conventional GDP fails to include important elements of the standard of living (e.g., environmental quality). The theoretical approach assesses welfare changes as the "indirect utility" changes that can be algebraically derived from the consumer demand function. Even if one subscribed to maximization of utility as a valid model of individual behaviour, this theoretical welfare measure is not viable in CGE models because of the aggregation problem discussed earlier. Neither measure addresses the distributional impacts associated with mitigation policies. We consider each of these issues in turn.

The flaws in the national accounts are well known (England, 1997). A recent study by the U.S. National Research Council made a number of recommendations for a sweeping overhaul of the national accounts to conform better to a reality that includes environmental and ecological values (Nordhaus and Kokkelenberg, 1999). In addition to the fact that the accounts "sometimes behave perversely with respect to environmental degradation and changing stocks of natural resources" and that they are "inconsistent in their treatment of different forms of wealth", the report notes that

> the conventional national accounts...give a very incomplete picture of the full scope of economic activity. By focusing only on marketed outputs and factors of production, the conventional accounts neglect a large number of economically significant inputs and outputs that are not bought and sold in markets. In the environmental area, these nonmarketed inputs and outputs often include the free goods and services provided by environmental assets such as air, water, forests, and complex ecosystems.... Because the conventional accounts omit such economically valuable but nonmarketed goods and services, they overstate the role of market inputs and outputs in economic welfare (pp. 25-26).

It is not necessary to accept the very large estimate of the value of such natural goods and services given by Costanza et al. (1997)[56] to realize that the omission of those services from the national accounts should send up a warning flag against unthinking use of measures such as GDP to describe economic well-being or the material standard of living[57].

More theoretically grounded measures for assessing welfare changes are plagued by the difficulties of the aggregation problem. A recent review article by Slesnick (1998) describes the current state of the art. While techniques exist that offer the possibility of theoretically consistent measures of welfare at the household level (if the difficulties of aggregating the utilities of members of the household into a single household utility function are assumed away)[58], there currently exists no theoretically consistent way of calculating changes in aggregate welfare. At the conclusion of his long survey, Slesnick states:

> While the conceptual issues of welfare measurement have been resolved at the micro [household] level, there is more ambiguity as to how to proceed to aggregate the welfare effects of households. Most agree that the procedure of assuming a representative agent and applying the techniques designed for households

---

56 Their estimate of the value of "natural services" is $ 33 trillion per year, which exceeds the global total annual value of goods and services that pass through markets.

57 The obvious flaws of GDP as welfare measure have prompted a number of practically minded groups to develop alternative measures. See e.g., the "Genuine Progress Indicator" promoted by the public policy group Redefining Progress (Stille, 2000). The "sustainable communities" movement also is concerned with developing indicators that better reflect quality of life and that can inform local and regional policy.

58 It should be noted that in the case of household welfare, comparisons between welfare levels of different price-quantity combinations can be made if enough information can be found on prices and quantities at the two different points being compared. In order to calculate the effects of hypothetical policy changes, stronger structural assumptions regarding functional forms of utility functions (for example) have to be made.

> tive agent and applying the techniques designed for households to aggregate data has serious shortcomings. Moreover...the use of income or other money-metric welfare indicators as arguments of a social welfare function also has many conceptual problems (p. 2159).

Economics has yet to find a way out of the dilemma posed by the impossibility of constructing a social welfare function that does not involve interpersonal comparisons of utility. Arrow's impossibility theorem (1951) remains an insurmountable obstacle to the construction of a non-dictatorial social choice function, and as Sen (1979) has pointed out, the use of a compensation principle to convert aggregate income changes into a Pareto-improving change is meaningless given that social mechanisms for such lump-sum transfers do not exist.

The reason that the lack of compensation mechanisms matters is, of course, that real economic policies always have distributional implications. Suppose a climate change mitigation policy leads to the loss of employment for some coal miners, increased gasoline prices at the pump, an expansion of business opportunities for the producers of cogeneration equipment and other energy-saving technologies, and a substitution of teleconferencing for a certain amount of business travel. In addition, the policy reduces climate risks, with corresponding changes in utility distributed across the population according to how those risks are valued, the spatial locations of the consumers (seashore residents vs. desert dwellers), etc. Is social welfare improved by the policy? Simple measurement (or prediction) of the GDP effect of the policies clearly cannot provide the answer. In addition to the fact that some of the most important effects are not priced, the effects of the policy have very different impacts on different individuals. In the absence of a mechanism whereby the "winners" from the policy could compensate the unemployed coal miners and others who stand to lose from it, there is no theoretically correct way to evaluate the welfare impact of the policy.

These distributional effects are a good part of what drives the politics of climate policy. In that sense, the outcome of the typical integrated assessment policy analysis is irrelevant. In the absence of some sort of compensation mechanism, even a potentially Pareto-improving climate policy would require virtual consensus to overcome the obstacles of special interest lobbying. The more politicized and salient the policy becomes, the less likely it is to command this kind of consensus, unless the risks and damages relative to the costs are so overwhelmingly large as to create near-unanimity in the populace[59].

Another reason climate policy is so difficult is that it involves issues of *intergenerational* distribution and equity. The long time lags between policy actions and their consequences mean that the costs and benefits of climate protection policies

---

59 An illustrative comparison is to the near-consensus on protection of the stratospheric ozone layer, in which the benefits of ozone layer protection, measured solely in terms of willingness to pay to avoid skin cancer risks, far outweighed the estimated costs of reducing dependence on ozone-depleting chemicals (Benedick, 1991, p. 63).

may be experienced by different generations entirely. The climate policy problem is not adequately represented by the modelling device of a social planner or central authority that is imagined to optimize aggregate welfare over an indefinitely long (or infinite) time horizon. This formulation of the problem implicitly involves interpersonal utility comparisons because the infinite time horizon encompasses the lives of successive generations of actual people. Furthermore, the undoubted fact that most living individuals prefer nearer-term consumption over the same consumption deferred to the future cannot legitimately be transformed into a discount rate applied to the infinite-horizon optimization problem. To do so entails making an implicit ethical judgment about the relative weights to be attached to the well-being of *different* individuals (who belong to different generations).

In an important paper that recasts the climate policy debate in an overlapping generations model (in which successive generations transact with each other through the "present" generation's rental of the capital stock accumulated by the "previous" generation), Howarth makes this ethical weighting of the different generations' welfare explicit. He demonstrates that the optimal degree of greenhouse gas control depends on the weights chosen; for a "utilitarian" weighting (in which each generation is given equal standing) the optimal level of control is more stringent than if future generations are given less weight than the current generation (Howarth 1998). This is only one example of a philosophically coherent approach; the economic literature on intergenerational justice and the "discounting problem" is extensive (see Lind, 1982, Cline 1992, Amano 1997 and Arrow et al. 1996 for an introduction)[60]. Regardless of the technical issues, however, the fundamental intergenerational equity problem remains unsolved. Technical economics can help frame the questions, but their resolution must be a matter of ethical choice[61].

There is one final point that should be made regarding the importance of distributional matters in climate policy evaluation. The working presumption of most models is that utility depends only on each individual's own level of income and consumption. This strong assumption is unlikely to be true in reality. Aside from the obvious interdependence of the utilities of household members, it is plausible to think that individuals' subjective well-being depends on the degree of inequality in society, the well-being of those with whom one is in social contact, the degree to which the society is perceived as conforming to notions of "fairness"[62], and so on. It is also known that the results of a welfare analysis change drastically if utilities are interdependent. For example, if social position enters the utility function, an externality results that leads to too much emphasis on acquisition of private goods as opposed to creation of public goods. Optimal policy would require intervention to

60 For very recent developments, see the downloadable papers presented at the conference on "Intergenerational Justice" held at the University of California, Davis, May 19-20, 2000 (http://www.econ.ucdavis.edu/seminars/ejsconf.html).

61 For an acute formulation of this inevitable choice see especially Sen (1982).

62 This is not to condone envy as a social value; concern for the well-being of the worst-off members of society is surely distinct from envy of the successful, and it is entirely possible to conceive of "fairness" in terms of the liberal ideals of freedom of contract and security of property rights. Needless to say, all existing societies fail in some ways to achieve "fairness," however defined.

correct this externality, and would suggest greater provision of public, non-excludable goods (including environmental protection of all types) that a laissez-faire system would produce (Howarth, 2000).

In addition to the direct effect on the equilibrium, this matters for climate policy because the correction of the greenhouse gas externality can have quite substantial wealth effects. The kinds of carbon taxes that would make a significant dent in emissions would raise tens of billions of dollars in the United States alone, enough to justify substantial reductions in other taxes if a principle of revenue neutrality were adhered to. Similarly, the creation of carbon emissions allowances would constitute a new form of property rights, the allocation of which could potentially have a noticeable effect on the distribution of wealth and income. The creation of a new form of wealth of such magnitude could affect preferences towards climate risks (because of the so-called "Engel curve" relating demand for environmental protection to wealth levels), and, if utilities are interdependent, the pattern of distribution of such substantial transfers could affect aggregate utility (however defined). Yet, the standard models treat none of these considerations.

2.3. IMPLICATIONS FOR CURRENT CLIMATE CHANGE POLICY ASSESSMENT

The above discussion captures a number of problems pertaining to neoclassical CGE analysis in general. We now illustrate how these and related shortcomings affect climate change mitigation assessment.

2.3.1. Mis-specifying energy services as an energy commodity

Standard models treat energy as a commodity that is consumed for its own sake. In these models, energy enters production and utility functions directly. But, in fact, energy may be more appropriately viewed as contributing to a service with multiple attributes. For example, energy can produce lighting, which also adds to comfort and aesthetics. Two problems arise from the misplaced categorization of energy as a commodity. One problem arises in the context of microeconomic choices, the other from the macroeconomic perspective.

Several writers have recognized that consumers do not derive utility from commodities per se, but from the commodities' attributes. Kelvin Lancaster, as one example, formulated utility arising from product characteristics (1966a, 1966b, 1971)[63]. This insight is highly relevant for the economic analysis of energy use. Energy is not consumed for its own sake, but because it yields a variety of services if combined with the appropriate appliances and capital equipment. The value of energy is embodied in both the level and quality of the multiple services that are provided. From the microeconomic perspective, if models assume only a single attribute of energy resources that is embodied in the price of energy, this tends to limit the role of other attributes in affecting consumer behaviour and choice. De-

63 For an overview of alternative approaches in consumption theory, see Goodwin, Ackerman, and Kiron (1997).

Canio and Laitner (1997), for example, demonstrate that if technology choice is modeled explicitly as a diffusion process subject to a range of influences, rather than a simple investment decision to be made on the basis of initial cost and energy bill savings, the penetration of new technologies is likely to be faster than indicated by conventional economic forecasts. Among the factors that tend to increase rates of market penetration, but that are not typically captured in standard models, are transmissions of more complete information about technology attributes, a growing consumer and business familiarity with the technologies, and the awareness of environmental impacts associated with the technologies.

From the macroeconomic perspective, modelling *energy services*, as opposed to merely *energy*, allows a finer depiction of how technology and inputs interact. If *energy* enters into the macroeconomic production function[64] *directly*, a reduction in energy use will reduce output in the nation's economy, all other things being equal. On the other hand, if *energy services* enter the macroeconomic production function, a reduction in energy need not be accompanied by a reduction in energy services, and hence, a reduction in output. On the contrary, when energy-efficiency investments provide energy bill savings and, often, other side benefits, a reduction in energy use can well be accompanied by an increase in both output and welfare.

2.3.2. *Unanticipated productivity benefits of energy efficiency investments*

In subsection 2.2.5 we argue that the literature on the "energy-efficiency gap" demonstrates that complete productive efficiency is not being reached in the provision of energy services. This same point applies to other inputs as well. Recent studies indicate, for example, that many business investments undertaken to reduce environmental impacts may also have productivity benefits beyond the environmental gains. This is especially true of investments that emphasize pollution prevention technologies and practices (Porter and van der Linde 1995; Florida, 1996). Similarly, investments in energy efficiency and renewable energy technologies may have other benefits not typically reflected in consumer or business decisions (Elliott et al., 1997; Romm, 1994, 1999). A review of 52 manufacturing case studies by Laitner and Finman (2000), for example, indicates that non-energy benefits of energy efficiency upgrades can be roughly equivalent to the energy bill savings. As a result, the average payback from these investments falls from four years when only energy savings are included in the analysis, to less than two years when both energy and non-energy savings are included.

The implications of failing to include these "co-benefits" within a modelling framework are significant in at least two ways. First, by failing to include total productivity gains within technology choice modules, any modelling exercise will necessarily under-represent the optimal adoption of energy efficiency and renewable energy technologies. Second, unless the co-benefits are explicitly represented in the various production functions of the models, the estimated economic impact of climate mitigation policies will be overly pessimistic. Moreover, a better representation of these

64 See section 2.2.7 above for difficulties with the concept of an aggregate production function.

co-benefits in future modelling exercises may have the added advantage of alerting businesses and consumers to these previously overlooked opportunities.

2.3.3. Voluntary programs and asymmetric information

Through voluntary programs that promote cost-effective energy efficiency and greenhouse gas emissions reductions, the U.S. Department of Energy (DOE) and the U.S. Environmental Protection Agency (EPA), among other federal agencies, have forged a series of partnerships with private and public organizations. These programs have demonstrated significant reductions in greenhouse gas emissions with net economic savings or very low costs. As but one example, the U.S. EPA's *ENERGY STAR*® and other voluntary programs exceeded their combined 1998 reduction goal of 14.6 million metric tons of carbon equivalent (MMTCE) by 16 percent. The agency's Climate Protection Division, the group that operates these programs, projects total reductions of more than 120 MMTCE by 2010 at current funding levels (Climate Protection Division, 1999). The documented success of EPA's voluntary programs prompted one group of researchers to conclude that such successes "cannot be reconciled with the view that energy-use decisions are made in efficient markets" (Howarth et al., 1999).

Despite these growing successes, conventional models tend to ignore or undervalue the successes of these marketing programs. For example, the Energy Information Administration credits voluntary programs operated under the U.S. Climate Change Action Plan with reductions of only 35-40 MMTCE (Energy Information Administration, 2000)[65]. The omission of the potential contributions of such programs, by definition, means that higher carbon charges will be required to achieve significant reductions in greenhouse gas emissions (Laitner, 1999b).

Luke notes that individuals and firms have a "natural tendency to choose from an *impoverished option bag*. Cognitive research in problem solving shows that individuals usually generate only about 30 percent of the total number of options on simple problems, and that, on average, individuals miss about 70 to 80 percent of the potential high quality alternatives" (Luke, 1998, page 114, emphasis in the original). Well-designed information programs can open up that "option bag" for both businesses and consumers. The evidence strongly indicates that voluntary programs induce firms to make investments in cost-saving technologies that firms have failed to exploit prior to the programs' implementation. Howarth et al. (1999) argue that the success of such programs is based on the ability to reduce market failures related to problems of asymmetric information and bounded rationality. More specifically, they propose that such limitations impair the effectiveness of both intra-firm organization and the coordination between equipment suppliers and their customers (see also Haddad et al., 1998; Haddad et al., 1999; and Paton, 1999a).

---

[65] Based on 1997 projections, the Climate Change Action Plan (CCAP) was expected to generate 96 MMTCE of energy-related carbon emission reductions by the year 2010. Hence, EIA is crediting CCAP actions with less than half of planned energy-related reductions. In addition, CCAP programs anticipate further reductions of 73 MMTCE from non-carbon emissions and land-use changes (Office of Global Change, 1997).

Discussing voluntary programs in the context of resource-based strategies, Paton (1999b) further suggests that such a framework "identifies at least seven types of opportunity for voluntary programs that neoclassical models can neither predict nor explain". These voluntary programs can: (1) challenge outdated routines, (2) accelerate process technology improvements, (3) stimulate product differentiation, (4) accelerate new product technologies, (5) eliminate collective action dilemmas, (6) overcome network externality problems, and (7) overcome retail market barriers. Paton concludes that firms participate in voluntary programs "for a variety of reasons not adequately captured in conventional theory" (pp. 1, 15-18). Among other things, the insights described here add reinforcement to the discussion on bounded rationality found in subsection 2.2.4.

3. Improving the current framework

The strengths of CGE models are clear: they accommodate detailed economic data and can trace indirect economic effects. If sufficiently disaggregated, they allow us to analyze the distribution of policy impacts over different groups of the population, depending on the extent of model detail. This capacity of CGE models has important relevance from both a political and economic perspective. Yet, as we have seen, both theory and evidence indicate a critical need to open up conventional models to better reflect the actual decision-making process and behaviour of both firms and consumers.

Providing a critical review of conventional models is not too difficult, but providing meaningful alternatives is a more daunting challenge. A tremendous amount of effort has gone into the development of these models and their parameterization. In the following, we suggest a series of strategies to address the weaknesses of conventional models we have discussed above, moving from interpreting and building on existing models to entirely new approaches for modelling economic activity, the development of which will take some time.

3.1. A SCHEMATIC ILLUSTRATION OF POLICY AND ORGANIZATIONAL PERSPECTIVES

As a first step in suggesting both near-term and long-term improvements to the modelling framework, we open the discussion with a schematic representation of many standard models. We then offer at least one alternative view of how such models might appear with an improved behavioural and organizational perspective. The intent is not to cover every aspect of the differences in modelling perspectives but to offer an illustration of critical elements that affect the modelling results.

Figure 2A corresponds to a conventional neoclassical production function – for example, of the constant elasticity of substitution (CES) type. The only real "policy handle" is an energy or carbon charge that enters the model as a change in the relative price of energy. This will clearly reduce energy consumption. But for reasons discussed above, it also tends to generate a pessimistic estimate of economic impact. This is not to say that a well-designed energy pricing policy (such as imposing a carbon tax with the revenues used to reduce taxes elsewhere in the economy)[66] cannot have a positive net effect, especially if integrated with other growth-enhancing policies[67]. Rather, it is to suggest that "price-auction" models[68] tend to understate the cost-effective opportunities to reduce greenhouse gas emissions. In this format, even with some form of human capital represented, the model strips from management many meaningful options in the selection of profit maximizing quantities of inputs and outputs (North 1990). The result is that such models tend to miss the behavioural and organizational response described by Paton (1999b), Hoffman (1997), DeCanio et al. (2000a), and others.

Figure 2A: A Schematic View of Energy Policy within Economic Models

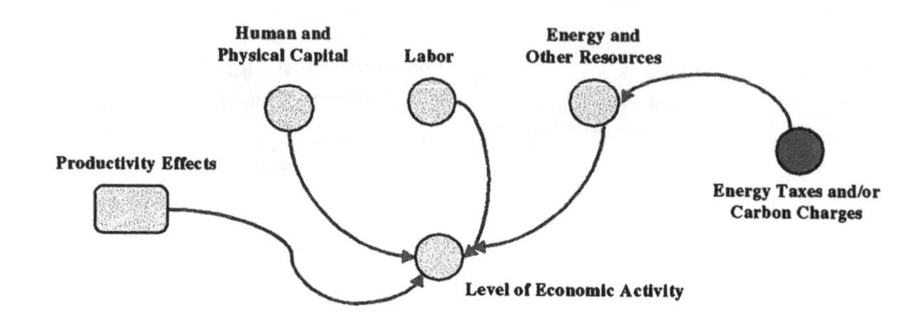

Figure 2B, on the other hand, offers one alternative illustration of how the dynamics of behavioural and organizational response, together with both price and non-price policies, might be captured within an economic assessment of mitigation strategies. The most immediate difference is that Figure 2B specifically reflects opportunities for such things as organizational change, human capital enhancement, and technology policy as they influence the pattern of energy services, productivity, and both economic output and environmental quality. Another important difference is to re-

---

66 See Parry and Bento (2000) and Jorgenson and Wilcoxen (1993) for evidence supporting the "strong double dividend hypothesis" that this kind of carbon tax revenue recycling could have a positive effect on economic activity independent of the environmental benefits. Even if the strong double dividend does not hold, it certainly must be the case that intelligent recycling of carbon tax revenues would greatly reduce any adverse economic effects of such a tax. (See also Brinner et al. (1991) which modeled three strategies for carbon tax recycling; one had positive impacts on growth in the long run).

67 Indeed, there is evidence that pricing signals that more accurately reflect both private and social costs, especially when combined with fundamental tax reform, can lower both carbon emissions and air pollution while supporting greater economic output at lower prices of non-energy goods and services (Norland et al., 1998).

68 That is, models in which the emissions reductions are accomplished either through imposition of a carbon tax or by means of the auction of tradable emissions permits.

flect the impact of price signals beyond energy consumption to include the influence on organizational response and the effectiveness of technology policy.

Figure 2B: A Schematic View of Climate Technology Policy and Organizational Change Perspectives within Emerging Models

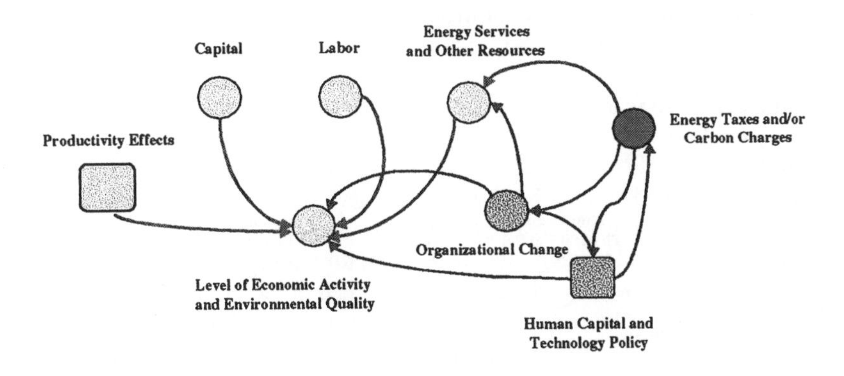

3.2. *SHORT-TERM IMPROVEMENTS IN CLIMATE POLICY MODELLING*

Porter has commented that in the area of business competition, most company leaders would not turn to economics for guiding insights. "Why the disconnect?" he asks. "Theories or models that require restrictive assumptions are untenable, because managers cannot hold everything else equal. Standard economic models of firms and product markets have captured little of the complexity and dynamism of actual competition. Managers are looking for ways of addressing important competitive questions that capture the complexities, rather than abstract from them" (Porter, 1998). In a similar way, consumers and policy makers are also looking for insights that capture the real-world complexities of climate change and associated mitigation strategies. This section describes several areas of short-term improvement as a first step in addressing such concerns.

3.2.1. *Interpretation of existing models*

Much can be done with small changes to existing models and their interpretation. The very first step is to recognize that the normative content of conventional models has no strong foundation, and that their results should be interpreted only as descriptive of observable economic variables (see Section 2.2.1) We can accept that CGE models capture some empirical regularities, but we have to concede that, in a world pervaded by externalities and distortions (some of them, for better or worse, politically intended!), with increasing returns, and with boundedly rational consumers and business firms, market outcomes do not represent efficient equilibria in any meaningful sense of the word "efficient". Moreover, market demand functions may be the sum of individual demand functions, but the theory of utility maximization is questionable at the individual level and it almost certainly does not apply to market aggregates (see Section 2.2.7). At best, market demand functions may offer a plausible description of what goes on in these markets absence any deep policy changes or other structural shifts. That is, they could be useful in describing what might happen to economic variables like sectoral output and employment, or to tax revenues and the trade balance, in response to not-too-drastic policy changes. Using the models responsibly for this purpose would require conducting careful sensitivity analyses with respect to parameter specifications.

North cautioned that once "we recognize that the costs of production are the sum of transformation and transaction costs, we require a new analytical framework of microeconomic theory" (1990, p. 28). Hence, a second step in the reinterpretation of conventional models may be to acknowledge that the economy operates interior to the production-possibilities frontier (as discussed in section 2.2.5). If so, conventional models should accept the value of bottom-up (engineering-economic) models as they point the way to decreasing the distance to the production-possibilities frontier[69]. Perhaps just as important is to come to terms with the many reasons why the economy operates within the frontier. From that point forward, mitigation assessments can more adequately reflect the impacts of both price and non-price emissions reduction policies.

3.2.2. *Accounting for informational and behavioural barriers*

In standard models, policies center around price signals as the leading mitigation strategy. However, there should be an explicit representation of policies other than price signals. Among the type of programs that can drive the desired investments are the previously described Energy Star initiatives operated by both the U.S. Environmental Protection Agency and the U.S. Department of Energy. Others programs include research and development subsidies, voluntary or negotiated agreements with business management, revenue-neutral "feebates" on the purchases of key energy-intensive equipment, cost-effective performance standards, tax incentives, and government adoption of energy efficiency measures in its own operations. These are all programs or incentives that, in effect, shift to the right the demand curve for efficiency and carbon-saving technologies. Furthermore, it should be kept in mind that the investment responses from the many businesses and households may be motivated by a variety of concerns such as climate change, energy security, and local energy-related pollution.

One possible representation of the influence of these programs or incentives is to estimate their impact on hurdle rates in the technology choice modules of energy/economic models. Typically, models adopt a price-based functional form, such as:

$$Energy_{new} = Energy_{base} * PRICE^{elasticity}$$

[69] For other reviews that discuss this perspective, see, Sanstad et al. (2000a); Laitner (1997); DeCanio (1997); and Koomey, et al. (1998).

where *PRICE* is an index of the energy price. But we can generate a functional form that responds to both price and policy or program signals, such as:

$$Energy_{new} = Energy_{base} * \left(\frac{PRICE_{new}}{HurdleRate_{new}} * \frac{HurdleRate_{base}}{PRICE_{base}} \right)^{elasticity}$$

where *HurdleRate* is a sector-specific value for both the base period and the new periods of economic activity[70]. To compare the impact of a price-led strategy with a combination of price and non-price programs and policies, let us assume that energy prices increase by 15 percent as a result of a $50/tonne carbon charge. To show how this will impact a projected consumption of 100 exajoules (EJ), let us further assume a price elasticity of -0.30. In that case we might have:

$$Energy_{new} = 100\ Exajoules * 1.15^{-0.30} = 95.9\ Exajoules$$

But we might also show the influence of a wide variety of programs and policies that, in effect, reduce the anticipated hurdle rate from 0.35 to 0.25. The many successes of the Energy Star programs reflect the capacity for this type of response (Climate Protection Division, 1999). Assuming a continued 15 percent price increase at the same elasticity, we would then have:

$$Energy_{new} = 100\ EJ * \left(\frac{1.15}{0.25} * \frac{0.35}{1.00} \right)^{-0.30} = 86.69\ EJ$$

In this example, we project a level of reduction beyond the price effect alone as a result of cost-effective voluntary programs. The impact of these various programs on the sector- or technology-specific hurdle rates can be approximated, econometrically calculated, or simply assumed – the representation is highly flexible. As a further illustration, let us posit that a 50 percent increase in the funding of such programs might further reduce in the hurdle rate to a value of 0.20. In this case, we would have the following combination of both price and non-price influences:

$$Energy_{new} = 100\ EJ * \left(\frac{1.15}{0.20} * \frac{0.35}{1.00} \right)^{-0.30} = 81.07\ EJ$$

70 The term *HurdleRate* is an implied "discount rate" that is actually a catch-all for all the factors affecting the return that firms require in order to undertake prospective investments. Thus, it incorporates the cost of capital (for projects of comparable risk) *plus* components reflecting informational asymmetries, organizational barriers, or remediable transactions costs that tend to impede investments. Experience shows that these non-financial barriers can be lowered by appropriate policy intervention (such as the provision of high-quality, credible information or centralization of the transactions costs) that allows policy simulations to be done by reduction of the hurdle rate. See DeCanio and Laitner (1997) for a full discussion. As an illustration in this case, let us assume the cost of capital is 10 percent. If a combination of risk premiums, information asymmetries, and other barriers are estimated to slow the adoption rate by the equivalent of another 25 percent in the implied discount rate, then we would say the *HurdleRate* for this example is equivalent to 35 percent.

In this illustration, the reduction of the hurdle rate from 0.35 to 0.20 – again as a result of successful program initiatives – together with a 15 percent price increase, has the same impact on energy consumption as a doubling of energy prices[71].

One further comment might be appropriate at this point. The example above assumed a constant price elasticity. Colander comments that "the concept of an unchanging elasticity of supply or demand is a product of a timeless model. In the real world, one would expect that the elasticity of supply and demand would change over time continually, becoming more elastic" (1992, p. 117). Consistent with that perspective, Laitner (1999c) outlined the movement of energy consumption elasticities in the United States with respect to both price and income. This analysis showed two things. The first was that both sets of elasticities changed significantly over time with price elasticity rising and then falling in the period 1973-1996. Second, and perhaps the surprising result, was that compared to the early 1970s, the income elasticity for energy consumption rose steadily. It now appears to show a greater magnitude of impact (in absolute terms) than does price elasticity. In other words, a 10 percent change in income will generate a much larger change in the demand for energy than will a 10 percent change in energy prices. Hence, one area of significant opportunity for economists is to explore the factors that cause elasticities to change over time, and then to incorporate such changes into mitigation assessments.

3.2.3. *Reflecting the multiple benefits of energy technology investments*

The return from an investment in a given energy technology is not limited only to direct energy consumption. In the industrial sector, for example, an energy efficiency upgrade may positively affect other steps in the production process. Fewer chemicals and other feedstocks may be consumed, or less time may be needed to get a product out the door. Often these secondary or co-benefits can generate as much value as the energy savings themselves. Yet, models that evaluate different technologies for their impact on energy consumption alone cannot estimate other costs or benefits unless such impacts are reflected in the assumptions of those models. As has already been suggested (in section 2.3.2), a better representation of co-benefits can affect both the mix of technologies that might be chosen by consumers or businesses and the impact on the larger economy as relative prices are changed in response to changed consumption patterns.

3.2.4. *Making use of scenario analysis*

Nations today face tremendous structural change and uncertainty that none of the existing policy models fully capture, especially as these uncertainties relate to climate strategies. This is true whether we are examining policy issues from the technological, market, institutional, or behavioural perspective. Yet, analysts can offer policymakers a way of thinking that helps organize the enormous volume of information relevant to complex policy issues. One important tool is scenario analysis, or what one publication refers to as "decision-focused scenario planning". (Global

---

71 The impact of a doubling of prices is 100 EJ * $2^{-0.30}$ = 81.22 EJ, comparable to the policy-induced reduction described in the text.

Business Network, 1993). Much of the work in scenario thinking builds on the work of Pierre Wack, a planner in the London Offices of Royal Dutch/Shell (Wack, 1985a, 1985b). Scenario analysis helped Shell respond quickly and successfully to the OPEC oil embargo following the Yom Kippur war in 1973.

The purpose of scenario planning is to explore several possible futures in a systematic way. One author describes scenarios as tools "for ordering one's perceptions about alternative future environments". The purpose of scenarios is not so much to produce an accurate picture of future events as to shape better decisions about the future. No matter how things might actually turn out, both the analyst and the policy maker will have "on the shelf" a scenario (or story) that resembles a given future and that will have helped them think through the opportunities and the consequences of that future. Such storytelling "resonates with what [people] already know, and leads them from that resonance to re-perceive the world". Scenarios, in other words, "open up your mind to policies [and choices] that you might not otherwise consider" (Schwartz, 1996).

The use of scenarios to examine either energy or climate policy is not new. Schipper and Meyers (1993), for example, developed a series of three scenarios of OECD sectoral energy intensities and how they might change through the year 2010 as a result of different policies. Koomey et al. (1998) described a number of technology-based scenarios for the U.S. energy system to understand what policies might induce both carbon savings and net economic benefits. At the same time, consultants at Arthur D. Little used scenario thinking to explore the influence of technological factors and societal values on companies seeking to sustain competitiveness in the facing of changing energy markets (Ross et al., 1997). Finally, Ruth et al. (2000) used scenario analysis to evaluate the impact of price-induced versus policy-induced climate mitigation strategies on the pulp and paper industry.

Other techniques can be used to complement decision-focused scenario planning. These include "deliberative polling" (Fishkin, 1995) and the forecast of emerging technologies using techniques such as environmental scanning and a modified Delphi survey (Halal et al., 1998). Deliberative opinion polling is a new form of public consultation in which ordinary citizens participate in a discussion and decisions about their energy future. In 1998, for example, a statistical sample of 250 customers of a major Texas utility were invited to spend a weekend in the deliberation of key issues related to electric utility service. Among the topics were the promotion of renewable energy and energy efficiency technologies, efforts to help low-income customers, and decisions about future investments in new fossil fuel plants (Center for Deliberative Polling, 1998).

Before the weekend, 51 percent of the sample believed the utility should use more renewable energy such as wind and solar power, and 48 percent wanted more energy efficiency programs. After the deliberations, 76 percent supported more renewable choices, and 64 percent wanted additional energy efficiency programs. Participants also were asked whether they would be willing to pay more on their monthly bills to support "electric generation using renewable technologies such as wind and solar power". Before the weekend, a majority were not willing to pay anything more.

After the weekend, however, a majority said they were willing to pay at least $6.50 a month more. Integrated Assessment focus groups consisting of laypersons, experts, policy makers, and representatives of business have also been organized in Europe, with a goal of combining social research and public participation (Querol et al., 1999; and Pahl-Wostl et al., 1998).

These approaches could significantly affect the design of both modelling exercises and policy proposals. Moreover, there may be a significant difference in outcomes as predicted by conventional models compared to a survey of experts who actively monitor market developments. For example, the George Washington University Forecast of Emerging Technologies projects that by 2010 or so most manufacturers will have adopted "green" methods in producing their consumer goods. According to this forecast, a significant portion of energy usage will be derived from renewable sources and biomass. In these years consumers will recycle about half their household waste. Around the year 2016, additional improvements in fossil fuel efficiency will reduce greenhouse gas emissions by one-half (Halal et al., 1998). Despite the fact that all of these possibilities would greatly mitigate greenhouse gas emissions, the current energy/economic models do not incorporate any of these emerging trends at the same scale or within the same time frames. To the extent that conventional models are more heavily weighted by historical data, they will fail to pick up newly emerging trends and technologies. All mitigation assessments are likely to be enhanced by incorporating these unconventional scenarios as complements to standard forecasting techniques. Consideration of a more inclusive set of technological possibilities widens the range of possible futures that can be contemplated.

3.3. LONGER TERM MODELLING ADVANCES

3.3.1. Pragmatic descriptions of aggregate behaviour

In our earlier discussion on the problem of aggregation, we noted that the "microfoundations" of aggregative neoclassical models are not really "micro" insofar as they do not depict the behaviour of individual agents. Rather, they use the representative agent device (see section 2.2.7). In the medium term, therefore, one could give up the neoclassical "microfoundations" and replace them with pragmatically formulated equations describing aggregate supply and demand in each market – a set of pragmatic "macro-functions", so to speak. This would avoid the normative efficiency implications of CGE analysis (Simon 1996). There are authors who have developed CGE models in this pragmatic spirit – among them Lance Taylor, Irma Adelman, and Sherman Robinson, working on economic development issues[72]. This branch of CGE modelling is in general less concerned with theoretical consistency and more with the usefulness of the models in capturing and representing real-world phenomena. While such work is to be welcomed, the question remains what exactly these pragmatic functions should look like. Empirical observation of aggregate market relationships may tell us something about their general shape, but will likely not provide enough data for their exact formulation.

---

72 See, for example, Adelman and Robinson (1978 and 1988); Lysy and Taylor (1980); and Taylor (1983 and 1990).

Pragmatic macro-functions could be derived from true micro-models, for example those that recognize interdependency of preferences, such as in micro-simulation models that are used in election result forecasting. Simon (1997) has offered a number of suggestions as to how economic research might move in this direction. It is entirely possible that the way to make progress will be to integrate elements of both CGE models and Multi-Agent Systems (Gallegati and Kirman, 1999; Peters and Brassel, 2000; Tesfatsion, 2000).

3.3.2. Incorporate multiple objective within models

Individual consumers and businesses juggle a variety of objectives or concerns as they choose their next investment or make their next purchase. The device of an all-encompassing utility function in neoclassical economics reduces these different objectives to a single function with multiple arguments. It may be more realistic to start from the premise that the individual agents seek to achieve specified levels of a number of variables, without assuming that they can order all possible combinations of desirable objectives or outcomes.

The development of multicriteria decision models can begin to embody this approach at the individual or firm level as well as for the economy as a whole. Multicriteria decision models could yield a more accurate description of individual choice (by consumers and managers), and thus improve the descriptive and predictive capability of models. If applied at the macro-level, multicriteria analysis may also serve to make explicit the variety of objectives, and their relative importance, which society as a whole may want to pursue. This would constitute a definite improvement over the common practice to regard some form of income or wealth maximization (e.g., GDP) as the sole societal objective.

One analytical tool that might provide further insights in this regard is "goal programming". Similar to the linear programming algorithm, goal programming solves for the set of choices that best meets multiple goals from among a variety of alternatives, all competing for limited resources[73]. It is different from linear programming in important ways, however. Where linear programming tries to achieve a single goal – for example, minimizing cost or maximizing profit – goal programming tries to achieve multiple goals. In effect, the goal programming model is based upon the "satisficing" principle first outlined by Simon in the 1950s[74]. This principle suggests that better decisions could be made if emphasis were given to achieving minimum levels of satisfaction for several goals rather than maximizing a single objective. Needless to say, this also differs from the optimizing behaviour that conventional models ascribe to economic agents. Some early work on the goal programming concept involved led to the development of a community-based energy management model (Lee and Laitner, 1986; Laitner and Kegel, 1988, 1989). Efforts are now underway to bring the issue forward in the energy and climate policy arenas (Laitner

---

73 For a more thorough discussion of the goal programming technique, see Render and Stair (1992), pages 307-320.

74 The reference to Simon's influence in the development of the goal programming model can be found in Lee (1976), pages 177-179.

and Hogan, 2000). The development of new modelling systems may lead to greater use of such models (Jones, et al., 1998).

3.3.3. *Research Agenda*

The existence of the kinds of defects in the conventional applied CGE models we have discussed suggests that there is considerable insight to be gained by basic economic research. On the theoretical side, progress appears to be possible in a number of specific areas. Long-term energy/economic forecasting models can be enhanced to incorporate the insights of the "new growth theory" of endogenous technological change. Models should be built that include increasing returns and other sources of positive feedback, giving rise to a variety of path-dependent outcomes that could illustrate a range of plausible future technology scenarios[75]. The heterogeneity of economic agents and their interaction should be explicitly modeled, thus providing true microfoundations for macro-models[76]. Production could be better represented in the models by a greater reliance on the modern theory of the firm. Intergenerational equity issues can be separated from purely technical discounting that reflects the marginal productivity of capital and the intertemporal preferences of individuals. The models can be modified to do a better job of taking "surprises" from the physical climate system into account (Sanstad et al., 2000b).

Additional empirical work remains to be done as well. The extent and exact causes of the deviations from neoclassical rationality at both the individual and organizational level are worthy of a great deal of exploration (Simon, 1997). Part of the appeal of the conventional approach is that economic maximization by rational agents gives structure to theorizing and provides sharp hypotheses that are subject to empirical testing. If the broader "behavioural" approach is to gain standing, the empirical implications of concepts like bounded rationality, co-evolution of tastes, the path-dependence of technological change, and the properties of psychologically-conditioned investment decisions need to be examined. Fortunately or unfortunately, the climate problem is of sufficiently long term duration that these research opportunities can be explored in depth.

4. Epilogue: Inventing the future

It should be clear that existing CGE models, as useful as they may be for some purposes[77]. do not and cannot provide the kind of precise policy guidance for which

---

[75] Work along these lines has been done at the Santa Fe Institute. See Anderson, Arrow, and Pines, (1988), and Arthur, Durlauf, and Lane (1997). Arthur (1994) contains some very innovative modelling examples.

[76] See Delli Gatti, Gallegati, and Kirman (eds.), 2000. See also Colander (2000).

[77] It should not be overlooked that CGE models and their empirical sub-structures provide useful information on how different sectors in the economy interact, on the identities that must be satisfied in any coherent system of national accounting, and on the historical trends that have characterized past economic activity. It is also the case that market conditions cannot be ignored in any economic model---traded commodities cannot long have prices differing by more than transportation costs in

they are typically invoked. That is, the existing theoretical and empirical framework of neoclassical economics does not offer the capability to determine accurately such things as the "cost" of alternative measures to reduce carbon emissions, the "benefits" of particular greenhouse gas abatement levels, or the "integrated assessment" of carbon control/tax/investment incentive policies. Over-reliance on the predictive power of even well specified equilibrium models leads to a specious precision that can mislead those who are unaware of the limitations of the models. More ominously, if their predictions are accepted uncritically, the models can be used to bolster the positions of special interest groups who know that they can elicit from the models the kinds of answers they desire simply by manipulation of the assumptions.

This does not mean, however, that the social and behavioural sciences cannot contribute significant insights to the climate policy debate. A number of the results we have discussed – regarding the aggregation problem, bounded rationality, the multiplicity of possible paths that can be taken by technological change – have the flavor of "impossibility theorems" that limit the kinds of conclusions that can be drawn on the basis of economic logic alone. These limits may preclude the specification of any model that can describe, with the kind of exactitude that enables physics to track the motions of the planets[78]. the development over time of the entire social system. But while it may be true that economics and the social sciences will *never* be able to offer predictive certainty regarding a phenomenon so complex as climate change, the models and techniques used for analysis can be improved along several dimensions. Another reason to be optimistic is that the range of predictive uncertainty attached to the projections of economic (and other socio-behavioural) models is wider than is ordinarily acknowledged. This lack of precision in forecasting is an opportunity as well as a limitation. It means that the range of options open to us may be wider than is commonly believed.

If that is the case, several implications follow. First, it is important for economics and the social sciences more generally to return *behaviour* to the center of the analytical stage. Instead of taking past responses to prices, past rates of technological progress, or past reactions to non-price policy initiatives as immutable guides to what will happen in the future, a treatment of the *behavioural assumptions* as part of the situation to be analyzed offers new theoretical and practical possibilities. Systematic thinking about the possible pathways that might lead from the present status quo to a desirable future is more in the spirit of "scenario analysis" (e.g., Global Business Network, 1993; and Ross et al., 1997) than predictive modelling, but it may also be more in keeping with the actual state of our understanding of the social process.

For example, if the evolution of tastes and technologies is path dependent, small nudges to the system at a time when major technological choices have not yet been

---

different locations, etc. As we will argue below, the challenge is to retain what is valid in the conventional approach while replacing the misleading and bias-inducing assumptions.

78 It is perhaps worth noting here that this apparent predictive power may be only a function of the time scale. Inter-planetary dynamics become chaotic after sufficient time (Wisdom, 1998; Duncan and Lissauer, 2000).

locked in can have strong repercussions in the future (Arthur, 1994). Setting off down one path rather than another may cause little or no reduction in the welfare of consumers or the profits of producers, now or in the future, but it can have very important consequences for the shape of things to come. This kind of argument for early action cannot be captured in equilibrium models in which the agents have rational expectations, perfect foresight, and the capacity to make long-lasting commitments (such as to a particular, allegedly "optimal" level of carbon emissions reductions at some time in the future when all current governments have served out their terms). Path dependence means that the choices made now are important, even if the exact nature of the future technological landscape cannot be known.

Including all relevant behaviour – whether personal, social, cultural, or organizational – in the set of factors that are potentially open to policy influence has its risks. There is always a temptation on the part of authorities and experts to substitute their own judgment and preferences for those of the people, and succumbing to this temptation can lead to the abuse of power. However, the danger cannot be avoided by an abdication of responsibility on the part of the governing elites. For long-term problems like climate change, doing nothing is as much a choice as action. Reliance on the ebb and flow of "politics as usual" to solve the problem is no guarantee of a good outcome. The problem of climate change is too complex to be addressed in political sound bites, and it is unrealistic to expect that effective policies can be discerned through polling data or by analyzing the reactions of unstructured focus groups. Even if the general public has a general predilection for environmental protection and for safeguarding the well-being of future generations, the public cannot be expected to have a command of all the technical and scientific issues relevant to policy design[79]. Solutions are not likely to appear spontaneously out of mass politics, but must be sought through a process in which expertise relies on peoples' best (rather than their basest) instincts and enlists their capacity for future-oriented thinking and planning. Just as individuals' utility functions have a social origin and exist in a social context, political attitudes are open-ended to some degree and susceptible to change in the face of new evidence and new arguments.

The wider range of possible outcomes from models that incorporate more of the reality of human and organizational behaviour implies that societies have considerable leeway in choosing the course of their future economic and technological development. Models can represent some kinds of choices well, and others partially, but no model can substitute for the responsibility to make those decisions that we are free to make. Social science can attempt to predict behaviour and its outcomes, but the predictions will always be contingent on the choices that are genuinely open. We have more control over the future than we might think, even if that control sometimes takes the negative form of inaction or blind stumbling. In the case of the climate, we can foresee the kind of future we would like our descendants to have – one

[79] And what if a majority (or significant minority) of the people are indifferent to the fate of future generations? Are those who can perceive the consequences of business as usual constrained from trying to educate their fellow-citizens about the impending danger? Such a position is inconsistent with democratic political theory, in which the most basic right (and obligation) of citizens is to engage in persuasion and to help develop policy through open discussion.

in which the benefits of a benign climate can be enjoyed by them as much as those benefits have been enjoyed by us. And once we have decided on the kind of future that we hope to achieve, social science – including an improved economics that draws on the insights of the full range of human experience – can help us bring it about.

5. References

Ackerman, Frank, 1997: Consumed in Theory: Alternative Perspectives on the Economics of Consumption, *Journal of Economic Issues,* Vol.31, No. 3 (September): pp. 651-664.

Ackerman, Frank, David Kiron, Neva R. Goodwin, Jonathan M. Harris, and Kevin Gallagher (eds.) 1997: *Human Well-Being and Economic Goals*. Washington, D.C.: Island Press.

Ackerman, Frank, 1999: Still Dead After All These Years: Interpreting the Failure of General Equilibrium Theory, Global Development and Environment Institute, Tufts University, Medford MA, manuscript.

Adelman, Irma and Sherman Robinson, 1978: *Income Distribution Policy in Developing Countries: A Case Study of Korea.* Stanford, CA: Stanford University Press.

Adelman, Irma and Sherman Robinson, 1988: Macroeconomic adjustment and income distribution: Alternative models in two economies, *Journal of Development Economics*, Vol.29: pp. 1-22.

Alchian, Armen, 1959: Costs and Output, in Moses Abramovitz et al., *The Allocation of Economic Resources: Essays in Honor of B.F. Haley*. Stanford: Stanford University Press: pp. 23-40.

Amano, Akihiro, 1997: On Some Integrated Assessment Modelling Debates, paper presented at the IPCC Asia-Pacific Workshop on Integrated Assessment Models, United Nations University, Tokyo, Japan, March 10-12, 1997.

Anderson, Elizabeth, 1993: *Value in Ethics and Economics*. Harvard University Press, 1993.

Anderson, Philip W., Kenneth J. Arrow, and David Pines (eds.), 1988: *The Economy as an Evolving Complex System.* Reading, Mass: Addison-Wesley Publishing Company.

Arrow, Kenneth J., 1951: *Social Choice and Individual Values*. New York: John Wiley & Sons.

Arrow, Kenneth J. and L. Hurwicz, 1958: On the Stability of Competitive Equilibrium I, *Econometrica*, Vol. 26: pp. 522-552.

Arrow, Kenneth J., H. D. Block, and L. Hurwicz, 1959: On the Stability of Competitive Equilibrium II, *Econometrica*, Vol. 27: pp. 82-109.

Arrow, Kenneth, 1962: The Economic Implications of Learning by Doing," *The Review of Economic Studies*, Vol. 29 (June): pp. 155-173.

Arrow, Kenneth, 1986: Rationality of Self and Others in an Economic System," *Journal of Business*, Vol.59, No.4, pt.2: pp. S385-S399.

Arrow, K. J., W. R. Cline, K-G. Maler, M. Munasinghe, R. Squitieri, and J. E. Stiglitz, 1996: Intertemporal Equity, Discounting, and Economic Efficiency, Chapter 4 of *Climate Change 1995: Economic and Social Dimensions of Climate Change*, Contribution of Working Group III to the Second Assessment Report of the Intergovernmental Panel on Climate Change, eds. Bruce, James P., Hoesung Lee, and Erik F. Haites. Cambridge: Cambridge University Press.

Arthur, W. Brian, Stephen N. Durlauf, and W.A. Lane (eds.), 1997: *The Economy as an Evolving Complex System II.* Reading, Mass: Addison-Wesley Publishing Company.

Arthur, W.B., 1994: *Increasing Returns and Path Dependence in the Economy.* Ann Arbor: The University of Michigan Press.

Asher, H., 1956: *Cost-Quantity Relationships in the Airframe Industry*, R-291. Santa Monica, CA: The RAND Corporation.

Barr, Nicholas, 1992: Economic Theory and the Welfare State: A Survey and Interpretation, *Journal of Economic Literature*, Vol.30 (June): pp. 741-803.

Benedick, Richard Eliot, 1991: *Ozone Diplomacy: New Directions in Safeguarding the Planet.* Cambridge, MA: Harvard University Press.

Bernow, Stephen, Irene Peters, Alexandr Rudkevich, Michael Ruth, 1998: *A Pragmatic CGE Model for Assessing the Influence of Model Structure and Assumptions in Climate Change Policy Analysis*, Boston, MA: Tellus Institute, November.

Bernow, Stephen, Karlynn Cory, William Dougherty, Max Duckworth, Sivan Kartha, Michael Ruth, 1999: *America's Global Warming Solutions*, Washington, DC: World Wildlife Fund.

Blinder, Alan.S., Elie R.D.Canetti, David E. Lebow, and Jeremy B. Rudd, 1998: *Asking about Prices. A New Approach to Understanding Price Stickiness.* New York: The Russell Sage Foundation.

Bohm, Peter, 1967: On the theory of 'second best', *The Review of Economic Studies,* Vol. 34: pp. 301-314.

Bovenberg, A. Lans, and Lawrence H. Goulder, 1996: Optimal environmental taxation in the presence of other taxes: general-equilibrium analyses, *American Economic Review*, Vol. 86, No. 4 (September): pp. 985-995.

Bovenberg, A. Lans and Lawrence H. Goulder, 1997: Costs of environmentally motivated taxes in the presence of other taxes: general equilibrium analyses, *National Tax Journal*, Vol. 50, No. 1 (March): pp. 59-87.

Bradsher, Keith, 1997: Government Studies New Risk on the Road, *The New York Times*, June 11: A14.

Bradsher, Keith, 2000: Ford Is Conceding S.U.V. Drawbacks, *The New York Times*, May 12: A1,C2.

Brinner, Roger E., Michael G. Shelby, Joyce M. Yanchar, and Alex Cristofaro, 1991: Optimizing Tax Strategies to Reduce Greenhouse Gases Without Curtailing Growth, The Energy Journal, Vol. 12 , No. 4: pp. 1-14.

Butler, D. J., 2000: Do non-expected utility choice patterns spring from hazy preferences? An experimental study of choice 'errors', *Journal of Economic Behaviour and Organization*, Vol. 41, No. 3 (March): pp. 277-297.

Cameron, Dr. Samuel, 2000: Nicotine addiction and cigarette consumption: a psycho-economic model," *Journal of Economic Behaviour and Organization*, Vol. 41, No. 3 (March): pp. 211-219.

Coase, R. H., 1960: The Problem of Social Cost, *Journal of Law and Economics*, Vol. 3: pp. 1-44.

Cowan, Robin, 1990: Nuclear Power Reactors: A Study in Technological Lock-in, *The Journal of Economic History*, Vol. L, No. 3: pp. 541-567.

Center for Deliberative Polling, 1998: Houston residents show dramatic shift in opinion about future energy sources during 'Deliberative Polling' event, a news release dated February 10, Department of Government at the University of Texas at Austin. URL: http://www.la.utexas.edu/research/delpol/index.html.

Charnes, A., W. Cooper, E. Rhodes, 1978: Measuring the Efficiency of Decision Making Units, *European Journal of Operations Research* Vol. 2: pp. 429-444.

Climate Protection Division, 1999: *Driving Investment in Energy Efficiency: Energy Star® and Other Voluntary Programs*. Washington, DC: U.S. Environmental Protection Agency, EPA-430-R-99-005, July.

Cline, William R., 1992: *The Economics of Global Warming*. Washington, DC: Institute for International Economics.

Colander, David, 1992: The Microeconomic Myth, in: Colander, David, and Reuven Brenner, eds., 1992. *Educating Economists*, Ann Arbor, MI: University of Michigan Press.

Colander, David, 1996a: (ed.). *Beyond Microfoundations. Post Walrasian Macroeconomics*. New York: Cambridge University Press.

Colander, David, 1996b: The macrofoundations of micro, in: *Beyond Microfoundations. Post Walrasian Macroeconomics*. Ed. by David Colander. New York: Cambridge University Press, pp.57-68.

Colander, David, 2000: New Millenium Economics: How Did It Get This Way, and What Way is It? *Journal of Economic Perspectives*, Vol. 14, No. 1 (Winter): pp. 121-132.

Conlisk, John, 1996: Why Bounded Rationality? *Journal of Economic Literature*, Vol. 34, No. 2 (June): pp. 669-700.

Cooper, W. et al., 1999: *Data Envelopment Analysis: A Comprehensive Text with Models, Applications, References, and DEA-Solver Software*. Dordrecht: Kluwer.

Costanza, Robert, Ralph d'Arge, Rudolf de Groot, Stephen Farber, Monica Grasso, Bruce Hannon, Karin Limburg, Shahid Naeem, Robert V. O'Neill, Jose Paruelo, Robert G. Raskin, Paul Sutton, and Marjan van den Belt, 1997: The value of the world's ecosystem services and natural capital, *Nature*, Vol. 387, No. 6630 (15 May): pp. 253-260.

Cox, David, and Richard Harris, 1985: Trade liberalization and industrial organization: Some estimates for Canada, Journal of Political Economy, Vol.93 (February): pp. 115-45.

Croson, Rachel. T. A., 2000: Thinking like a game theorist: factors affecting the frequency of equilibrium play, *Journal of Economic Behaviour and Organization*, Vol. 41, No. 3 (March): pp. 299-314.

Daly, Herman E. and John B. Cobb, Jr., 1989: *For the Common Good: Redirecting the Economy Toward Community, the Environment, and a Sustainable Future*. Boston, MA: Beacon Press.

Debreu, Gérard, 1970: Economies with a Finite Set of Equilibria, *Econometrica* Vol. 38: pp. 387-92.

Debreu, Gérard, 1974: Excess demand functions," *Journal of Mathematical Economics*, Vol.1: pp. 15-23.

DeCanio, Stephen J., 1993: Barriers within firms to energy-efficient investments, *Energy Policy*, Vol. 21, No. 9 (September): pp. 906-914.

DeCanio, Stephen J., 1994: Energy Efficiency and Managerial Performance: Improving Profitability While Reducing Greenhouse Gas Emissions, in David L. Feldman, ed., *Global Climate Change and Public Policy*. Chicago: Nelson-Hall Publishers.

DeCanio, Stephen J., 1999: Estimating the Non-Environmental Consequences of Greenhouse Gas Reductions Is Harder Than You Think, *Contemporary Economic Policy*, Vol. 17, No. 3 (July): pp. 279-295.

DeCanio, Stephen J. and John A. "Skip" Laitner, 1997: Modelling Technological Change in Energy Demand Forecasting: A Generalized Approach," *Technological Forecasting and Social Change*, Volume 55 (October): pp. 249-263.

DeCanio, Stephen J., Catherine Dibble, and Keyvan Amir-Atefi, 2000a: The Importance of Organizational Structure for the Adoption of Innovations, *Management Science* (forthcoming).

DeCanio, Stephen J., Catherine Dibble, and Keyvan Amir-Atefi, 2000b: Organizational Structure and the Behaviour of Firms: Implications for Integrated Assessment, Climatic Change (forthcoming).

DeCanio, Stephen J., William E. Watkins, Glenn Mitchell, Keyvan Amir-Atefi, and Catherine Dibble, 2000c: Complexity in Organizations: Consequences for Climate Policy Analysis, in *Advances in the Economics of Environmental Resources*, eds. Richard B. Howarth and Darwin C. Hall. Greenwich, CT: JAI Press Inc. (forthcoming).

De Lang, J. Bradford, Andrei Shleifer, Lawrence Summers, and Robert Waldmann, 1990: Noise Trader Risk in Financial Markets, Journal of Political Economy, Vol.98: pp. 703-38.

Delli Gatti, Domenico, Mauro Gallegati, and Alan P. Kirman, 2000: The economics of heterogeneity, in; Delli Gatti, Gallegati, Kirman (eds.), 2000: *Interaction and Market Structure*. Springer.

Delli Gatti, Domenico, Mauro Gallegati, and Alan P. Kirman (eds.), 2000: *Interaction and Market Structure. Essays on Heterogeneity in Economics*. Berlin, Heidelberg, New York: Springer.

Dennis, Michael L., E. Jonathan Soderstrom, Walter S. Koncinski, Jr., and Betty Cavanaugh, 1990: Effective Dissemination of Energy-Related Information, *American Psychologist*, Vol. 45, No. 10 (October): pp. 1109-1117.

Deutsch, David, 1997: *The Fabric of Reality*. Allen Lane The Penguin Press, Penguin Books Ltd.

Duncan, Martin J. and Jack J. Lissauer, 2000: Solar System Dynamics, in *Encyclopedia of the Solar System*, http://www.apnet.com/solar/Contents/chap34.htm.

Elliott, R. Neal, Miriam Pye, and John A. "Skip" Laitner, 1997: Considerations in the Estimation of Costs and Benefits of Industrial Energy Efficiency Projects. *Proceedings of the 32nd Annual Intersociety Energy Conversion Engineering Congress*, Paper #97-551, Honolulu, HI, July.

Energy Information Administration, 2000: *Assumptions to Annual Energy Outlook 2000. With Projections to 2020*. Washington, DC: U.S. Department of Energy, DOE/EIA-0554 (2000), January.

Energy Innovations, 1997: *Energy Innovations: A Prosperous Path to a Clean Environment*. Washington, DC: Alliance to Save Energy, American Council for an Energy-Efficient Economy, Natural Resources Defense Council, Tellus Institute, and Union of Concerned Scientists.

England, Richard W., 1997: Alternatives to Gross National Product: A Critical Survey, in: *Human Well-Being and Economic Goals*. Ed. by Frank Ackerman, David Kiron, Neva R. Goodwin, Jonathan M. Harris, and Kevin Gallagher. Washington, D.C.: Island Press: pp. 373-402.

Etzioni, Amitai and Paul R. Lawrence (eds.), 1991: *Socio-Economics: Toward a New Synthesis*, Armonk, NY: M.E. Sharpe.

Evans, Leonard, 1985: Car Size and Safety: Results From Analyzing U.S. Accident Data, in Proceedings of the Tenth International Technical Conference on Experimental Safety Vehicles, Oxford, England, July 1-4.

Fan, Chinn-Ping, 2000: Teaching children cooperation---An application of experimental game theory, *Journal of Economic Behaviour and Organization*, Vol. 41, No. 3 (March): pp. 191-209.

Fisher, Franklin M., 1968a: Embodied technology and the existence of labor and output aggregates, *Review of Economic Studies,* Vol.35, No.4. Reprinted in Fisher (1993), pp.69-102.

Fisher, Franklin M., 1968b: Embodied technology and the aggregation of fixed and movable capital goods, *Review of Economic Studies,* Vol.35, No.4. Reprinted in Fisher (1993), pp.103-120.

Fisher, Franklin M., 1969: The existence of aggregate production functions, *Econometrica,* Vol. 37, No. 4. Reprinted in Fisher (1993), pp.1-30.

Fisher, Franklin M., 1982: Aggregate production functions revisited: The mobility of capital and the rigidity of thought, *Review of Economic Studies,* Vol. 49, No. 4. Reprinted in Fisher (1993).

Fisher, Franklin M., 1983: *Disequilibrium foundations of equilibrium economics*. Cambridge: Cambridge University Press.

Fisher, Franklin M., 1989: Adjustment Processes and Stability, in: *The New Palgrave: General Equilibrium.* Ed. by John Eatwell, Murray Milgate and Peter Newman. First American edition. New York and London: W.W. Norton. pp. 36-42.

Fisher, Franklin M., 1993: *Aggregation: Aggregate Production Functions and Related Topics*. Cambridge, MA: MIT Press.

Fishkin, James S., 1995: *The Voice of the People: Public Opinion and Democracy.* Yale University Press.

Florida, Richard, 1996: Lean and Green: The move to environmentally conscious manufacturing, California Management Review, (39(1): pp. 80-105.

Friedman, Milton, 1953: *Essays in Positive Economics*. Chicago: University of Chicago Press.

Gabaix, Xavier and David Laibson, 2000: A Boundedly Rational Decision Algorithm, *American Economic Review (Papers and Proceedings of the One Hundred Twelfth Annual Meeting of the American Economic Association)*, Vol. 90, No. 2 (May): pp. 433-438.

Gallegati, Mauro and Alan P. Kirman (eds.), 1999: *Beyond the Representative Agent*. Northampton, MA: Edward Elgar Publishing.

Geller, Howard, Stephen Bernow, and William Dougherty, 1999: *Meeting America's Kyoto Protocol Target: Policies and Impacts*, Washington, DC: American Council for an Energy-Efficient Economy, November.

Ginsburgh, V. and M. Keyzer, 1997: *The Structure of Applied General Equilibrium Models*. Cambridge, MA: MIT Press.

Global Business Network, 1993: Customer 20/20: Breaking the Future Trap — Volume 1 Assuring Future Customer Options through Scenario Planning, Palo Alto, CA: Electric Power Research Institute, TR-101694, V.1, January.

Goodwin, Neva, Frank Ackerman, and David Kiron (eds.), 1997: *The Consumer Society*. Washington, D.C.: Island Press.

Goulder, Lawrence H., 1995: Effects of Carbon Taxes in an Economy with Prior Tax Distortions: An Intertemporal General Equilibrium Analysis, *Journal of Environmental Economics and Management* (October).

Goulder, Lawrence H. and Stephen H. Schneider, 1999: Induced Technological Change and the Attractiveness of CO2 Abatement Policies, *Resource and Energy Economics,* Vol. 21, Nos. 3-4 (August): pp. 211-253.

Grandmont, Jean-Michel, 1998: Expectations Formation and Stability of Large Socioeconomic Systems, *Econometrica*, Vol. 66, No. 4 (July): pp. 741-781.

Haddad, Brent M., Richard B. Howarth, and Bruce Paton, 1998: Energy Efficiency and the Theory of the Firm, Proceedings of the 1998 ACEEE Summer Study on Energy Efficiency in Buildings, American Council for an Energy-Efficient Economy, Washington, DC, August.

Haddad, Brent M., Richard B. Howarth, and Joan Brunkard, 1999: Why Firms Participate in Energy Star: Mechanisms and Initial Data, unpublished manuscript.

Hahn, Frank, 1982: Stability, in: *Handbook of Mathematical Economics.* Ed. by Kenneth J. Arrow and Michael D. Intriligator. Amsterdam: North-Holland. Vol. II, pp. 745-793.

Halal, William E., Michael D. Kull, and Ann Leffmann, 1998: The George Washington University forecast of emerging technologies," *Technological Forecasting & Social Change*, Vol. 59, No. 1 (September), p89(22).

Hanson, Donald A. and John A. "Skip" Laitner, 1999: Investment in Energy-Efficient Technology in an Economic Growth Context, a presentation to the Eastern Economics Association Meetings, Boston, MA, March 12-14.

Hanson, Donald A. and John A. "Skip" Laitner, 2000: An Economic Growth Model with Investment, Energy Savings, and CO2 Reductions, presented to Salt Lake City meetings of the Air & Waste Management Association, June 18-22.

Hanson, Donald, 1999: *A Framework for Economic Impact Analysis and Industry Growth Assessment: Description of the AMIGA System*, Argonne National Laboratory, Argonne, IL, April.

Harris, Richard G., 1984: Applied general equilibrium analysis of small open economies with scale economies and imperfect competition, *American Economic Review*, Vol.74 (December): pp. 1016-31.

Helpman, Elhanan and Paul R. Krugman, 1985: *Market Structure and Foreign Trade*. Cambridge, MA: M.I.T. Press.

Herings, P. Jean-Jaques, 1996: *Static and Dynamic Aspects of General Disequilibrium Theory*. Boston, London, Dordrecht: Kluwer Academic Publishers.

Herings, P. Jean-Jacques, 1997: A globally and universally stable price adjustment process, *Journal of Mathematical Economics*, Vol.27: pp. 163-193.

Hildenbrand, Werner, 1994: *Market Demand: Theory and Empirical Evidence*. Princeton, N.J.: Princeton University Press.

Hoerner, J. Andrew and Avery P. Gilbert, 1999: *Assessing Tax Incentives for Clean Energy Technologies: A Survey of Experts Approach,* Working Paper 3, Washington, DC: Center for a Sustainable Economy, November.

Horowitz, John. K. and K. E. McConnell, 2000: Values elicited from open-ended real experiments, *Journal of Economic Behaviour and Organization*, Vol. 41, No. 3 (March): pp. 221-237.

Howarth, Richard B., 1997: Energy Efficiency and Economic Growth, *Contemporary Economic Policy*, Vol. 14, No. 4 (October): pp. 1-9.

Howarth, Richard B., 1998: An Overlapping Generations Model of Climate-Economy Interactions, *Scandinavian Journal of Economics*, Vol. 100, No. 3: pp. 575-591.

Howarth, Richard B., 2000: Climate Change and Relative Consumption," in *Advances in the Economics of Environmental Resources*, eds. Richard B. Howarth and Darwin C. Hall. Greenwich, CT: JAI Press Inc. (forthcoming).

Howarth, Richard B., Brent M. Haddad, and Bruce Paton, 1999: The Economics of Energy Efficiency: Insights from Voluntary Participation Programs, *Energy Policy* (forthcoming).

Humphrey, Steven J., 2000: The common consequence effect: testing a unified explanation of recent mixed evidence, *Journal of Economic Behaviour and Organization*, Vol. 41, No. 3 (March): pp. 239-262.

Ingrao, Bruna and Giorgio Israel, 1990: *The Invisible Hand. Economic Equilibrium in the History of Science*. Cambridge, Mass: MIT Press. (Published in Italian as *La Mano Invisibile*. Roma, Bari: Gius. Laterza & Figli Spa, 1987).

Interlaboratory Working Group on Energy-Efficient and Clean-Energy Technologies, 2000: (forthcoming). *Scenarios for a Clean Energy Future*, prepared for Office of Energy Efficiency and Renewable Energy, U.S. Department of Energy.

Interlaboratory Working Group, 1997: Scenarios of U.S. Carbon Reductions: Potential Impacts of Energy-Efficient and Low-Carbon Technologies by 2010 and Beyond. Berkeley, CA: Lawrence Berkeley National Laboratory and Oak Ridge, TN: Oak Ridge National Laboratory. LBNL-40533 and ORNL-444 (September). URL address: http://www.ornl.gov/ORNL/Energy_Eff/CON444.

Jacoby, Henry D., Richard S. Eckaus, A. Denny Ellerman, Ronald G. Prinn, David M. Reiner, and Zili Yang, 1997: CO_2 Emission Limits: Economic Adjustments and the Distribution of Burdens, *The Energy Journal*, Vol. 18 No. 3: pp. 31-58.

Janssen, Maarten C.W., 1993: *Microfoundations. A Critical Inquiry.* London: Routledge.

Jerison, M., 1984: Social Welfare and the Unrepresentative Representative Consumer, Discussion Paper, State University of New York at Albany, NY.

Jones, D. F., M. Tamiz, and S. K. Mirrazavi, 1998: Intelligent solution and analysis of goal programmes: The GPSYS system, *Decision Support Systems* 23: pp. 329-332.

Jones, Charles I. and John C. Williams, 1998: Measuring the Social Return to R&D. *Quarterly Journal of Economics* Vol. 113, No. 4 (November): pp. 1119-1135.

Jorgenson, Dale W., and Peter J. Wilcoxen, 1993: Reducing U.S. Carbon Emissions: An Econometric General Equilibrium Assessment, *Resource and Energy Economics*, Vol. 15: pp. 7-25.

Kahnemann, Daniel, and Amos Tversky, 1979: Prospect Theory: An Analysis of Decision under Risk, *Econometrica*, Vol. 47, No. 2: pp. 263-91.

Kahnemann, Daniel, Paul Slovic, and Amos Tversky (eds.), 1982: *Judgment under Uncertainty: Heuristics and Biases.* Cambridge, U.K., and New York: Cambridge University Press.

Kehoe, Patrick J. and Timothy J. Kehoe (eds.), 1995: *Modelling North American Economic Integration.* Boston: Kluwer Academic Publishers.

Kehoe, Timothy J., 1985: Multiplicity of Equilibria and Comparative Statics, *The Quarterly Journal of Economics*, Vol.99: pp. 119-147.

Kehoe, Timothy J., 1998: Uniqueness and Stability, in: *Elements of General Equilibrium Analysis*, ed. by Alan P. Kirman: pp. 38-87.

Kehoe, Timothy J., D.K. Levine, and P.M. Romer, 1992: On Characterizing Equilibria of Economies with Externalities and Taxes as Solutions to Optimization Problems, *Economic Theory*, Vol.2: pp. 43-68.

Kirman, Alan P., 1989: The intrinsic limits of modern economic theory: The emperor has no clothes, *The Economic Journal,* Vol.99 (Conference Supplement): pp. 126-139.

Kirman, Alan P., 1992: Whom or What Does the Representative Individual Represent? *Journal of Economic Perspectives*, Vol.6 No.2 (Spring): pp. 117-136.

Kirman, Alan P., ed. 1998: *Elements of General Equilibrium Analysis.* Oxford, U.K., and Cambridge, Mass.: Blackwell Publishers.

Kirman, Alan P., 1998: Introduction, in Kirman (ed.), *Elements of General Equilibrium Analysis.* Oxford, U.K., and Cambridge, Mass.: Blackwell Publishers, pp. 1-9.

Kline, David M. and John A. "Skip" Laitner, 1999: Policies to Enhance Technology Diffusion and Market Transformation," A paper presented at the annual conference of the International Association of Energy Economics in Florida. September.

Koomey, Jonathan G., R. Cooper Richey, "Skip" Laitner, Robert J. Markel, and Chris Marnay, Technology and Greenhouse Gas Emissions: an Integrated Scenario Analysis Using the LBNL-NEMS Model, EPA 430-R-98-021, U.S. Environmental Protection Agency, Washington, DC, September 1998. http://enduse.lbl.gov/Projects/GHGcosts.html.

Krause, Florentin, Jonathan Koomey, and David Oliver, 1999: *Cutting Carbon Emissions While Making Money: Climate Saving Energy Strategies for the European Union* (Executive Summary). El Cerrito, CA: International Project for Sustainable Energy Paths.

Krugman, Paul R., 1979: Increasing Returns, Monopolistic Competition, and International Trade, *Journal of International Economics*, Vol. 9, No. 4 (November): pp. 469-479.

Krugman, Paul R., 1980: Scale Economies, Product Differentiation, and the Pattern of Trade, *American Economic Review*, Vol. 70: pp. 950-959.

Laitner, John A. "Skip" and Hodayah Finman, 2000: Productivity Benefits from Industrial Energy Efficiency Investments. A monograph prepared for the EPA Office of Atmospheric Programs, Washington, DC, March.

Laitner, John A. "Skip", 1997: WYMIWYG (What You Measure is What You Get): The Benefits of Technology Investment as a Climate Change Policy, a paper given to the 18th Annual North American Conference of the USAEE/IAEE, San Francisco, CA. September 7-10, 1997.

Laitner, John A. "Skip", 1999a: The Economic Effect of Climate Policies that Increase Investments in Cost-Effective, Energy-Efficient Technologies, Presented at the 74th Annual Western Economic Association International Conference, San Diego, CA, July.

Laitner, John A. "Skip", 1999b: The Impact of Climate Policies on the Cost of Carbon Reductions in the United States, A working paper prepared for the EPA Office of Atmospheric Programs, Washington, DC, July.

Laitner, John A. "Skip", 1999c: Engineering-Economic Analyses of GHG Emissions Reductions: Microeconomic Foundations with Macroeconomic Results, *Proceedings of The IEA International Workshop on Technologies to Reduce Greenhouse Gas Emissions: Engineering-Economic Analyses of Conserved Energy and Carbon*, Washington, DC, May 5-7.

Laitner, John A. "Skip" and Jack Kegel, 1988: Community Energy Choices: The Goal Programming Concept As An Economic Development Assessment Tool for Utility/Community Based Energy Management Programs, *Proceedings of the Sixth Annual Conference of the National Association of Regulatory Utility Commissioners*, Columbus, OH, September.

Laitner, John A. "Skip" and Jack Kegel, 1989: Evaluating Community Energy Management Strategies Using the OPTIONS Model. *Proceedings of the 1989 Energy Program Evaluation Conference*. Chicago, IL. August.

Laitner, John A. "Skip" and Kathleen Hogan, 2000 (forthcoming): Solving for Multiple Objectives: The Use of the Goal Programming Model to Evaluate Energy Policy Options, a paper for the ACEEE Buildings Summer Study, August 2000.

Laitner, John A. "Skip", Kathleen Hogan, and Donald Hanson, 1999: Technology and Greenhouse Gas Emissions: An Integrated Analysis of Policies that Increase Investments in Cost Effective Energy-Efficient Technologies, *Proceedings of the Electric Utilities Environment Conference*, Tucson, AZ, January.

Lancaster, Kelvin, 1966 a: A New Approach to Consumer Theory, *Journal of Political Economy*, Vol.74 (April): pp. 132-157.

Lancaster, Kelvin, 1996 b: Change and Innovation in the Technology of Consumption, *American Economic Review*, Vol. 56 (May): pp. 14-23.

Lancaster, Kelvin, 1971: *Consumer Demand: A New Approach*, Columbia University Press.

Lee, Sang M., 1976: *Linear Optimization for Management*. New York, NY: Mason Charter Publishers.

Lee, Sang M., Skip Laitner, and Yung M. Yu, 1986: A Goal Programming Decision Support System for Community Energy Management Programs, a paper presented to the 1986 Meeting of the Decision Science Institute, Honolulu, HI, November.

Lind, Robert C., 1982: A Primer on the Major Issues Relating to the Discount Rate for Evaluating National Energy Options, in: Robert C. Lind (ed.), 1982. *Discounting for Time and Risk in Energy Policy*. Washington, D.C.: Resources for the Future. pp. 21-94.

Lipsey, R.G. and Kelvin Lancaster, 1957: The general theory of second best," *The Review of Economic Studies*, Vol. 24: pp. 11-32.

Loewenstein, George, 2000: Emotions in Economic Theory and Economic Behaviour, *American Economic Review (Papers and Proceedings of the One Hundred Twelfth Annual Meeting of the American Economic Association)*, Vol. 90, No. 2 (May): pp. 426-432.

Lucas, Robert E., Jr. 1988: On the Mechanics of Economic Development. *Journal of Monetary Economics* 22 (July): pp. 3-42.

Luke, Jeffrey S., 1998: *Catalytic Leadership: Strategies for an Interconnected World,* San Francisco, CA: Jossey-Bass Publishers.

Lysy, F. J., and L. Taylor, 1980: The general equilibrium income distribution model, in: L. Taylor, E. Bacha, E. Cardoso, and F. J. Lysy, eds. *Models of Growth and Distribution for Brazil*. London: Oxford University Press.

Manne, Alan S., and Richard R. Richels, 1992: *Buying Greenhouse Insurance: The Economic Costs of Carbon Dioxide Emission Limits*. Cambridge, Mass.: MIT Press.

Mantel, R., 1974: On the characterisation of aggregate excess demand, *Journal of Economic Theory*, Vol. 7: pp. 348-353.

Martel, Robert J., 1996: Heterogeneity, Aggregation, and a Meaningful Macroeconomics, in David Colander, ed., *Beyond Microfoundations: Post Walrasian Macroeconomics*. New York: Cambridge University Press. pp. 127-144.

Martino, Joseph P., 1999: The environment for technological change, *International Journal of Technology Management*, Vol. 18 (Sept.-Oct.) i1-2 p4(7).

McKee, Michael and Edwin G. West, 1981: The theory of second best: A solution in search of a problem, *Economic Inquiry,* Vol.19 (July): pp. 436-448.

Morishima, M., 1984: The good and bad uses of mathematics, in *Economics in Disarray,* ed. P. Wiles and G. North. Oxford: Basil Blackwell.

National Academy of Sciences, 1992: *Policy Implications of Greenhouse Warming: Mitigation, Adaptation, and the Science Base*. Washington, DC: National Academy Press.

Nelson, Richard R., 1995: Recent Evolutionary Theorizing About Economic Change, *Journal of Economic Literature*, Vol. 33, No. 1 (March): pp. 48-90.

Nelson, Richard R., 1996: *The Sources of Economic Growth,* Cambridge, MA: Harvard University Press.

Nelson, R.R., & Winter, S.G., 1982: *An evolutionary theory of economic change,* Cambridge, MA: Belknap Press.

Nordhaus, William D. and Edward C. Kokkelenberg, (eds.), 1999: *Nature's Numbers: Expanding the National Economic Accounts to Include the Environment*. Washington, DC: National Academy Press.

Norland, Douglas L., Kim Y. Ninassi, and Dale Jorgenson, 1998: *Price it Right: Energy Pricing and Fundamental Tax Reform* (Executive Summary), Washington, DC: Alliance to Save Energy.

North, Douglass C., 1990: *Institutions, Institutional Change and Economic Performance*, New York, NY: Cambridge University Press.

Office of Global Change, 1997: *Climate Action Report: 1997 Submission of the United States of America Under the United Nations Framework Convention on Climate Change*, Washington, DC: U.S. Department of State, July.

Office of Technology Assessment (OTA), U.S. Congress, 1991: *Changing by Degrees: Steps to Reduce Greenhouse Gases*, OTA-0-482. Washington, DC: U.S. Government Printing Office.

Olson, Mancur, 1965: *The Logic of Collective Action: Public Goods and the Theory of Groups*. Cambridge, MA: Harvard University Press.

Pahl-Wostl, Claudia, Carlo C. Jaeger, Steve Rayner, Christoph Schär, Marjolein van Asselt, Dieter M. Imboden, and Andrej Vckovski, 1998: Regional Integrated Assessment and the Problem of Indeterminacy, in Peter Cebon, Urs Dahinden, Huw Davies, Dieter M. Imboden, and Carlo C. Jaeger, eds., *Views from the Alps: Regional Perspectives on Climate Change*. Cambridge, MA: The M. I. T. Press.

Parry, Ian W. H. and A.M. Bento, 2000: Tax deductions, environmental policy, and the 'double dividend' hypothesis, *Journal of Environmental Economics and Management*, Vol. 39 No. 1 (January): pp. 67-96.

Paton, Bruce, 1999a: Resources as Capital: Insights on Efficiency Gains from Voluntary Environmental Policies, a working paper for the European Research Network on Voluntary Approaches (CAVA). http://www.akf.dk/cava/wp.htm.

Paton, Bruce, 1999b: Voluntary Environmental Initiatives and Sustainable Industry, Greening of Industry Network Conference, Chapel Hill, NC, November.

Peters, Irene, Frank Ackerman, and Stephen Bernow, 1999: Economic theory and climate change, *Energy Policy,* Vol.27, No. 9 (September): pp. 501-504.

Peters, Irene and Kai-H. Brassel, 2000: Integrating Computable General Equilibrium Models and Multi-Agent Systems – Why and How, in: *2000 AI, Simulation and Planning in High Autonomy Systems.* Ed. by Hessam S. Sarjoughian, Francois E. Cellier, Michael M. Marefat, and Jerzy W. Rozenblit. Simulation Councils, Inc.: pp. 27-35.

Porter, Michael E., 1991: America's Green Strategy: Environmental Standards and Competitiveness, *Scientific American*, Vol. 264, No. 4 (April): 168.

Porter, Michael E. and Claas van der Linde, 1995: Toward a New Conception of the Environment-Competitiveness Relationship," *Journal of Economic Perspectives*, Vol. 9, No. 4 (Fall): pp. 97-118.

Porter, Michael E., 1998: The Adam Smith address: location, clusters, and the 'new' microeconomics of competition, *Business Economics*, Vol. 33, No. 1 (Jan.): p7(7).

Querol, Cristina, Åsa Gerger, Bernd Kasemir, and David Tàbara, 1999: Citizens' Recommendations for Addressing Climate Change: A Participatory Integrated Assessment Exercise in Europe, ULYSSES WP-99-4, Darmstadt University of Technology, Center for Interdisciplinary Studies in Technology, Darmstadt, Germany.

Ravetz, Jerry, 1999: Models as Metaphors, ULYSSES WP-99-3, Darmstadt University of Technology, Center for Interdisciplinary Studies in Technology, Darmstadt, Germany.

Render, Barry and Ralph M. Stair, Jr., 1992: *Introduction to Management Science*, Boston, MA: Allyn and Bacon.

Repetto, Robert and Duncan Austin, 1997: *The Costs of Climate Protection: A Guide for the Perplexed*. Washington, DC: World Resources Institute.

Rizvi, S. Abu Turab, 1994: The microfoundations project in general equilibrium theory, *Cambridge Journal of Economics,* Vol.18: pp. 357-377.

Romer, Paul M., 1986: Increasing Returns and Long-Run Growth. *Journal of Political Economy* Vol. 94 (October): pp. 1002-1037.

Romer, Paul M., 1990: Endogenous Technological Change. *Journal of Political Economy*, Vol. 98, pt 2: pp. S71-S102.

Romer, Paul M., 2000: Thinking and Feeling, *American Economic Review (Papers and Proceedings of the One Hundred Twelfth Annual Meeting of the American Economic Association)*, Vol. 90, No. 2 (May): pp. 439-443.

Romm, Joseph J., 1994: *Lean and Clean Management: How to Boost Profits and Productivity by Reducing Pollution*. New York/Tokyo/London: Kodansha International.

Romm, Joseph J., 1999: *Cool Companies: How the Best Businesses Boost Profits and Productivity by Cutting Greenhouse Gas Emissions*. Washington, DC: Island Press.

Ross, E.H. Christopher, J. Ladd Greeno, and Albert Sherman, 1997: Scenario Thinking: Planning for the Futures You Want (and the Futures You Just Might Get), *Prism*, Third Quarter.

Rubinstein, Ariel, 1998: *Modelling Bounded Rationality*. Cambridge, MA: The M. I. T. Press.

Rust, John, 1997: Dealing with the Complexity of Economic Calculations, paper for workshop on Fundamental Limits to Knowledge in Economics, Santa Fe Institute, August 3, 1996, Santa Fe, revised October 1997.

Ruth, Matthias, Brynhildur Davidsdottir, and John A. "Skip" Laitner, forthcoming. Using Climate Change Policies to Promote Efficiency in the US Pulp and Paper Industry, *Energy Policy*.

Saari, Donald G., 1985: Iterative Price Mechanisms, *Econometrica,* Vol.53, No.5 (September): pp. 1117-1131.

Saari, Donald G. and C.P. Simon, 1978: Effective Price Mechanisms, *Econometrica,* Vol. 46: pp. 1097-1125.

Saari, Donald G. and S.R. Williams, 1986: On the Local Convergence of Economic Mechanisms, *Journal of Economic Theory,* Vol.40, No.1 (October): pp. 152-167.

Sagoff, Mark, 1981: At the Shrine of Our Lady Fatima, or Why Political Questions Are Not All Economic, *Arizona Law Review*, Vol.23: pp. 1283-1298.

Sagoff, Mark, 1988: Some Problems with Environmental Economics, *Environmental Ethics*, Vol.10, No.1: pp. 55-74.

Sagoff, Mark, 1994: Should Preferences Count? *Land Economics*, Vol.2 (May): pp. 127-145.

Sanstad, Alan, 1999: Endogenous Technological Change and the Crowding Out Problem in Climate Policy, paper presented at the Conference on Economics and Integrated Assessment, Pew Center on Global Climate Change, Washington DC (July 21-22).

Sanstad, Alan H., Stephen J. DeCanio, and Gale A. Boyd, 2000a: Estimating Bounds on the Macroeconomic Effects of the CEF Policy Scenarios, Appendix E-4 of *Scenarios for a Clean Energy Future*, prepared by the Interlaboratory Working Group on Energy-Efficient and Clean-Energy Technologies. Washington, DC: U.S. Department of Energy, Office of Energy Efficiency and Renewable Energy.

Sanstad, Alan H., Stephen J. DeCanio, Richard Howarth, Stephen Schneider, and Starley Thompson, 2000b: *New Directions in the Economics and Integrated Assessment of Global Climate Change*. Washington, DC: Pew Center on Global Climate Change.

Schipper, Lee, and Stephen Meyers, 1993: Using scenarios to explore future energy demand in industrialized countries, *Energy Policy*, Vol. 23 (March): pp. 264-275.

Schumpeter, Joseph A., 1942: *Capitalism, Socialism and Democracy.* 2nd ed., New York: Harper Brothers.

Schwartz, Peter, 1996: *The Art of the Long View: Planning for the Future in an Uncertain World.* New York, NY: Doubleday.

Sen, Amartya K., 1979: The Welfare Basis of Real Income Comparisons: A Survey, *Journal of Economic Literature*, Vol. 17, No. 1: pp. 1-45.

Sen, Amartya K., 1982: Approaches to the Choice of Discount Rates for Social Benefit-Cost Analysis, in: Robert C. Lind (ed.), 1982. *Discounting for Time and Risk in Energy Policy.* Washington, D.C.: Resources for the Future. pp. 325-376.

Sen, Amartya K., 1985: Well-Being, Agency and Freedom: The Dewey Lectures 1984, *Journal of Philosophy*, Vol.82: pp. 169-221.

Shackley, S. & Darier, É., 1998: Seduction of the Sirens: global climate change and modelling, *Science and Public Policy*, Vol. 25, No. 5: pp. 313-325.

Shafer, Wayne, and Hugo Sonnenschein, 1982: Market demand and excess demand functions, in *Handbook of Mathematical Economics,* eds. Kenneth J. Arrow and Michael D. Intriligator. Amsterdam: North-Holland. Vol. II, pp. 671-693.

Shoven, John B. and John Whalley, 1992: *Applying General Equilibrium.* New York: Cambridge University Press.

Simon, Herbert A., 1996: Economic Rationality: Adaptive Artifice. In: *The Sciences of the Artificial.* 3rd edition. Cambridge, MA: MIT Press, pp. 25-49.

Simon, Herbert A., 1996: *Models of Bounded Rationality.* Vol. 3. *Empirically Grounded Economic Reason.* Cambridge, MA: MIT Press.

Simon, Herbert A., 1947: *Administrative Behaviour*. New York: Macmillan, 1947.

Simon, Herbert A., 1957: *Models of Man.* New York: Wiley.

Simon, Herbert A., 1963: Discussion, *American Economic Review*, Vol. 53, No. 2: pp. 229-231.

Simon, Herbert A., 1997: *An Empirically Based Microeconomics.* Cambridge University Press.

Simon, Herbert A.,. *Models of Bounded Rationality.* Vol.1 *Economic Analysis and Public Policy.* Cambridge, MA: MIT Press.

Simon, Herbert A.,. *Models of Bounded Rationality*. Vol.2 *Behavioural Economics and Business Organization.* Cambridge, MA: MIT Press.

Slesnick, Daniel T., 1998: Empirical Approaches to the Measurement of Welfare, *Journal of Economic Literature*, Vol. 36, No. 4 (December): pp. 2108-2165.

Sonnenschein, Hugo, 1972: Market excess demand functions, *Econometrica,* Vol.40: pp. 549-563.

Sonnenschein, Hugo, 1973: Do Walras identity and continuity characterize the class of community excess demand functions?, *Journal of Economic Theory,* Vol.6: pp. 345-354.

Starmer, Chris, 2000: Developments in Non-Expected Utility Theory: The Hunt for a Descriptive Theory of Choice under Risk, *Journal of Economic Literature*, Vol. 38, No. 2 (June): pp. 332-382.

Stern, Paul C., and Elliot Aronson (eds.), 1984: *Energy Use: The Human Dimension*, Committee on Behavioural and Social Aspects of Energy Consumption and Production, National Research Council, New York, NY: W.H. Freeman and Company.

Stern, Paul C., 1986: Blind Spots in Policy Analysis: What Economics Doesn't Say About Energy Use, *Journal of Policy Analysis and Management*, Volume 5, No. 2: pp. 200-227.

Stern, Paul C., 1992: What Psychology Knows About Energy Conservation, *American Psychologist,* Vol. 47, No. 10 (October): pp. 1224-1232.

Stern, Paul C., and Gerald T. Gardner, 1981: Psychological Research and Energy Policy, *American Psychologist*, Vol. 36, No. 4 (April): pp. 329-342.

Stille, Alexander, 2000: A Happiness Index With a Long Reach, *The New York Times* (May 20): A17.

Stoker, Thomas M., 1993: Empirical approaches to the problem of aggregation over individuals, *Journal of Economic Literature,* Vol. 31 (December): pp. 1827-1874.

Taylor, Lance, ed., 1990: *Socially Relevant Policy Analysis. Structuralist Computable General Equilibrium Models for the Developing World.* MIT Press, Cambridge, MA.

Taylor, Lance, 1983: *Structuralist Macroeconomics: Applicable Models for the Third World.* New York: Basic Books.

Tisdell, Clem, 1996: *Bounded Rationality and Economic Evolution. A Contribution to Decision Making, Economics and Management.* Cheltenham, U.K.: Edward Elgar.

Tellus Institute, 1997: *Policies and Measures to Reduce CO_2 Emissions in the United States: An Analysis of Options for 2005 and 2010.* Boston, MA: Tellus Institute.

Tesfatsion, Leigh, 2000: Website Agent-Based Computational Economics, www.econ.iastate.edu/tesfatsi/ace.htm.

Thaler, Richard H., 2000: From Homo Economicus to Homo Sapiens, *Journal of Economic Perspectives*, Vol. 14, No. 1 (Winter): pp. 133-141.

The Economist 1999 Dec 18.: Rethinking Thinking, pp. 69-71.

Tobin, James, 1986: The future of Keynesian economics, *Eastern Economic Journal,* Vol.13, No.4.

Union of Concerned Scientists and Tellus Institute, 1998. *A Small Price to Pay: US Action to Curb Global Warming Is Feasible and Affordable.* Cambridge, MA: UCS Publications.

Van DeVeer, Donald, and Christine Pierce (eds.), 1998: *The Environmental Ethics and Policy Book. Philosophy, Ecology, Economics.* 2nd ed. Belmont, CA: Wadsworth Publishing Company.

Van der Laan, G. and A. J. J. Talman, 1987: A Convergent Price Adjustment Process, *Economics Letters*, Vol.23: pp. 119-123.

Vatn, Arild and Daniel W. Bromley, 1994: Choices without Prices without Apologies, *Journal of Environmental Economics and Management*, Vol.26: pp. 129-148.

Villar, A., 1996: *General Equilibrium with Increasing Returns.* Berlin: Springer.

Wack, Pierre. 1985a: The Gentle Art of Reperceiving–Scenarios: Uncharted Waters Ahead (part 1 of a two-part article), *Harvard Business Review*, (September-October): 73-89.

Wack, Pierre, 1985b: The Gentle Art of Reperceiving–Scenarios: Shooting the Rapids (part 2 of a two-part article), *Harvard Business Review*, (November-December): pp. 2-14.

Weidlich, Wolfgang and Günter Haag, 1983: *Concepts and Models of a Quantitative Sociology: The Dynamics of Interacting Populations.* Berlin, New York: Springer.

Weintraub, E. Roy, 1979: *Microfoundations: The Compatibility of Microeconomics and Macroeconomics.* Cambridge, U.K.: Cambridge University Press.

Weitzman, Martin L., 1974: Prices vs. quantities, *The Review of Economic Studies,* Vol. 41 (October): pp. 447-491.

Weyant, John, Henry Jacoby, James Edmonds, and Richard Richels, (eds.), 1999: *The Costs of the Kyoto Protocol: A Multi-Model Evaluation*, a special issue of the *Energy Journal*, Spring.

Williamson, Oliver E., 1992: Transaction Cost Economics and Organization Theory, as found in, Giovanni, Dosi, David J. Teece, and Josef Chytry, *Technology, Organization, and Competitiveness: Perspectives on Industrial and Corporate Change,* New York, NY: Oxford University Press.

Wisdom, J., 1998: A Brief Introduction to Chaos in the Solar System, http://geosys.mit.edu/~solar/text/short.html.

Wright, T. P., 1936: Factors Affecting the Cost of Airplanes, *Journal of the Aeronautical Sciences*, 3: pp. 122-128.

Yang, Zili, et al., 1996: *The MIT Emissions Prediction and Policy Assessment (EPPA) Model.* MIT Joint Program on the Science and Policy of Global Change, Report No. 6. Cambridge, MA: MIT.

Zizzo, Daniel John, Stephanie Stolarz-Fantino, Julie Wen, and Edmund Fantino, 2000: A violation of the monotonicity axiom: experimental evidence on the conjunction fallacy, *Journal Of Economic Behaviour And Organization*, Vol. 41, No. 3 (March): pp. 263-276.

Acknowledgement

The authors gratefully acknowledge the advice and comments from the participants at the March 20-21, 2000 workshop in Karlsruhe, Germany. In addition, we would like to especially acknowledge the contributions of Philippe Crabbé, Frank Ackerman, Brent Haddad and other reviewers who helped us strengthen and tighten the logic and arguments of the paper. Needless to say, any mistakes or omissions remain our own responsibility.

Cultural Discourses in the Global Climate Change Debate

Steven Ney
Michael Thompson

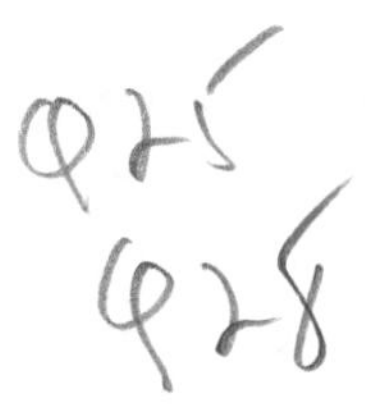

1. Introduction

If Paul Sabatier and Hank Jenkins-Smith (1987, 1993) are right, we should have witnessed significant policy movement in the global climate change debate by now. These two political scientists contend that any sustained and meaningful change in policy, regardless of the particular policy area, takes at least 10 years. Although roughly a decade[1] has passed since global climate change arrived on the policy agendas of most advanced industrial countries, any observer would be hard-pressed to say with any conviction that much has changed since the Earth Summit in Rio, in 1992.

Perhaps this is a little unfair: there has, indeed, been change. The legacy of Rio and its aftermath (the summits in Berlin and Kyoto as well as the multitude of conferences, symposia, and workshops at virtually all societal and administrative levels) has been climate change strategies, policy positions, carbon-dioxide emission targets, reduction policies, energy efficiency concepts, and a plethora of models for greening virtually all areas of human activity at international, national, and even local levels. Yet, at the same time, the actual carbon emissions have continued to increase at an alarming pace. In short and in policy jargon, while we have seen an enormous growth in *policy output*, there has been little in the way of *policy outcomes.*

In situations such as these, political scientists and policy analysts are called upon (or, more realistically, feel the irresistible urge) to explain why the growth in policy seems to have had little impact on the "policy problem". Explanations, and models to tailor these explanations, abound. We may think about the difficulties of implementing and co-ordinating policy at various levels of governance (including the transnational): states will be states and will maximize their national interest. Alternatively, we may look for faults in the relevant decision-making processes: policy processes may not be sufficiently "rational" to provide the necessary decisions for combating climate change. We may even attribute failure to "powerful interests" that block "rational" policy.

However, while these contentions are undoubtedly true, they often reveal an explicit or implicit distaste for *politics*: here, politics is perceived as something distinctly unsavoury. Politics (whether in the guise of "national interest", "bounded rationality", "the implementation gap", or "non-generalizable interests") often appears as

1 Eight years if you count from the Rio Earth Summit, in 1992; fifteen years from the Villach Conference, in 1985, where the IPCC (Intergovernmental Panel on Climate Change first saw the light of day.

E. Jochem et al. (eds.), Society, Behaviour, and Climate Change Mitigation, 65–92.

the culprit: the insurmountable hurdle to "better and more rational" policy. In general, this perception takes two forms: "the spirit is willing but the flesh is weak" (the argument that good policy is spoiled by partisan implementation), or, conversely, the "healthy body blighted by madness' (decision-making processes are crippled by particular interests). Either way, the implication is that, if only we could divorce policy-making (rational, pure, sanitary) from politics (irrational, messy, stained), then we would be in the position to solve the problem[2].

Let us, for a moment, assume that politics and policy-making are irredeemably intertwined. Let us further assume that politics – the wheeling and dealing, the persuading and exhorting, the conflicts and alliances, as well as the claiming and blaming – is not only an important part of policy-making but also a characteristic feature of human sociability. How, then, could we explain the performance of global climate change policy?

In this paper, we will look at ways of understanding the climate change debate in terms of its inherently political characteristics. In particular, we will explore how divergent socio-institutional discourses frame and define policy issues (such as global climate change). As a result, these discourses structure policy-debates and thus determine both policy outputs and outcomes. Understanding policy-making means understanding how these socio-institutional discourses set up political conflict, provide channels for political alliances and rivalries, and supply the means for policy solutions.

In essence, this paper is a summary of several pieces of work (Schwarz and Thompson 1985, Rayner and Thompson 1998; Rayner, Thompson and Ney 1998; Thompson and Ney 1999). In Section 1, we look at why it is important to analyse rhetoric, argument, and discourse. This section also provides an abridged framework for understanding the social origins of policy arguments or policy stories. Section 2 outlines three such socio-institutional framings (or stories) of global climate change, each of which provides a cogent account of the problem, its causes and its solution. In Section 3, we briefly outline what such a socio-institutional analysis means for policy-making on global climate change, and in Section 4 we provide a glimpse, in terms of *scenario planning*, of the sorts of policy tools that this politics-accepting approach calls for.

2. Policy Arguments and Policy Stories

Why are cultural discourses important?

Conventional models of the policy-process would have us believe that policy-making is a rational activity. Policy-makers identify and select issues, filter out some and promote others to the agenda, review all possible solutions, decide on the most efficient policy response, and, finally, implement this response. While such a model

---

2 In fairness, political scientists and policy analysts, at least on the academic circuit, have grown increasingly suspicious of policy science's promise of a rational policy process (cf. Fischer and Forester 1993; Dryzek 1990; Parsons 1995; Stone 1998; Hood 1998).

(usually referred to as a "stagist" or "synoptically rational" approach) is undoubtedly of value when thinking about policy processes in the abstract, it tells us very little about the overt and covert conflicts, the shifting alliances, and the creative uses of knowledge and facts: in short, the politics we observe in real-life policy-making.

- Policy processes, apart from being rational approaches to solution design, are also social processes. As such, politics and policy-making are also about the purposive manipulation and deployment of symbols, claiming and blaming, persuasion and communication.
- Policy-making does not take place in a social vacuum but rather emerges from a highly complex web of social relations. Regardless of whether these are the relationships between individual politicians, within interest groups, between party members, or across partly divisions, policy actors are situated in a trellis of social ties that make up the political system. The social networks both constrain and facilitate political action. On the one hand, policy actors are limited by the formal and informal rules (e.g. the "whips"); on the other hand, it is precisely these social structures that make policy action possible (Smith 1993).
- Policy-making, then, is a process based on shared values and norms that emerge from the social interaction of policy actors. These systematically and symbolically structured sets of ideas provide policy actors with the means to understand and make sense of policy events: policy actors evaluate political events by referring to these shared ideas, values and symbols. Perception of policy issues is thus filtered through the different "perceptual lenses" (Allison 1971) provided by social relations. What is to count as political or non-political, as fact or value, as a key issue or a non-issue, is not an objective reality "out there". Rather, the significance of any political event, any particular issue, or even any political structure emerges from policy actor' interpretations of political reality. In short, policy actors socially construct the political world in which they operate: running the maze *and* building it, in contrast to the behaviourist's rats that simply run the maze they are put in. What is more, policy actors will use these social constructions to exhort, cajole, and persuade potential allies as well as to antagonize, scandalize, and intimidate political rivals.

In this politics-accepting view, policy-making is an inherently communicative endeavour that follows a different logic to the rational stage models. Rather than each individual rational actor pursuing goals by his or her strategically manipulating other policy actors, this view suggests that policy action is based, at least in part, on communication. Communication is based on reasoned argument, in which both actors draw on intersubjectively shared norms and values, with the aim of establishing mutual understanding as the basis for policy action (Habermas 1987; Eriksen and Weigard 1997). Communicative and symbolic resources are thus important elements of the policy process, and the realization poses the question: how can we go about analysing them?

What is a Policy Argument?

If communication, persuasion and the use of symbolism is an integral part of the policy-process, understanding policy-making implies looking at its argumentative aspects. Contemporary theorists of the policy process, such as John Dryzek, Bruce Jennings, Giandomenico Majone, Frank Fischer and John Forester, have pointed to the argumentative rhetorical and justificatory content of much of policy-making. Policy formulation, policy planning, and even policy implementation, they maintain, emerge from argumentative processes that conventional policy analysis has so far ignored. Paying attention to these communicative processes means taking seriously "the actual performances of argumentation and the practical rhetorical work of framing analyses, articulating them, [and] constructing senses of value and significance" (Fischer and Forester 1993:5).

Rather than understanding policy-makers as problem-solvers who apply scientific methods to cure society of its ills, theorists of the "argumentative turn" suggest we think of policy-makers as performers who seek to persuade an audience. In order to convince other policy-makers and the public, participants in the policy process use political symbols to construct credible and persuasive policy arguments.

A policy argument tells a story: it provides a setting, points to the heroes and villains, follows a plot, suggests a solution, and, most importantly, is guided by a moral. Since policy arguments are designed to persuade, they are necessarily value-oriented (Adler and Haas 1992:p. 29); Rein and Schoen 1994). Yet this does not mean that policy arguments are mere opinion. Policy arguments explicate problems by recourse to rational methods: logic, consistency, and objectivity play an important role.

Significantly, an argumentative analysis of the policy process views all appeals to truth and objectivity in terms of argumentative performance: policy arguments are successful not because they are based on an objective standard, but because they persuade. Of course, the fact that some policy arguments are based on a method that is widely viewed as credible may itself be compelling: economic forecasts based on sophisticated econometric models are at present more plausible than financial predictions based on astrological star-charts.

The policy argument approach looks at the effects of discourse on policy-making. In doing so, it introduces both a reflexive and a critical element into policy analysis. Focusing on the rhetorical performance of the policy argument enables the analyst to step back from substantial policy issues and discern how and why a policy argument accrues credibility: that is, how discourse affects the public debate. It allows us to understand why certain types of policy argument are marginalized and why others achieve dominance: a policy argument that can muster sufficient levels of credibility will be able to dominate a policy debate.

Once the notion of credibility is thematized, the analyst can raise issues of political legitimacy: this is the element of criticism in argumentative analysis. The argumentative approach recognizes that credible policy arguments are not necessarily legitimate: rationality and objectivity are not sufficient conditions for a policy argument to secure credibility, nor are they always necessary. Credibility is not an absolute quality of a policy argument: it depends on the rhetorical performance of policy-actors, the internal logic of the policy storyline, the normative orientations of the policy audience, and actual power relations in the public sphere.

The policy argument as a unit of analysis allows us to scrutinize both the cognitively rational (meaning objective) and communicatively rational (meaning normative) components of any single policy argument. The policy argument approach implies that every policy story not only gives us an interpretation of the "facts" concerning any given issue complex but is also guided, implicitly or explicitly, by a particular vision of the world: policy arguments always follow a moral agenda.

The Origins of Policy Stories

Policy stories or policy arguments, however, do not materialize from thin air. Nor do policy actors "invent" them like an author invents a piece of fiction. Rather, policy stories emerge from the socio-institutional structures that make up the web of social relations in the policy-making process.

How, then, can we think about these structures and their associated policy stories? Firstly, we can think about the extent to which social relations are patterned by external rules and regulations. The more a particular set of social relations is subject to rules, norms, and regulation, the smaller the ability of individuals to negotiate and bargain their relations with others. Secondly, we can think about social relations in terms of external barriers. If relations are patterned so as to define a strong group, then individual members will have little scope to determine the form of their relationships with those who are not within that group. These two conditions, which are the basis for what is called *grid-group analysis* (Douglas 1978) provide us with four fundamental forms of social solidarity.

- If social relations are relatively free from rules and regulation (low grid, that is) in addition to providing very weak external boundaries (low group) then individuals will negotiate and bargain their social relations according to their perceived utility. Social scientists refer to this sort of solidarity – a setting in which individuals interact on the basis of free negotiation and contract, and where policy outcomes result from the self-interested interactions of policy actors – as a *market* (an essentially *individualist* form of solidarity).

- Conversely, if sequences of relationships "bend around" to define closed loops rather than the ever-proliferating open networks that characterize markets, and if the resultant groupings are then arranged into some sort of rank order (high group, in other words, and high grid), then we will have the diametric opposite of individualist solidarity. Regulation will be strong (rendering symmetrical transactions difficult) and boundaries will be marked (leaving little scope for negotiation between individuals). In short, this form of social solidarity, as social scientists have long been aware, exhibits *hierarchical* characteristics. Individuals interact on the basis of clearly defined roles and tasks, policy outcomes are the result of careful, rational management and propriety – who has the right to do what, and to whom – is an omnipresent concern.

- When sequences of relations bend around on themselves (high group) but the resultant groupings preserve symmetry by not entering into ranked relationships with one another (low grid) we have a form of solidarity that provides an alternative to both markets and hierarchies. This "enclavist" or sect-like form of solidarity, which has been rather neglected by social science, is labelled *egalitarianism*. Here, individual choice in the forming of social relations is relatively free from rules with the group, but is severely limited beyond the boundary. Policy outcomes here emerge, not from a process of bargaining (market), nor from rational management (hierarchy), but from the consensual interactions of equal individuals: "direct consent", as it is sometimes called.
- Finally, when rules restrict the individual's ability to form social relationships (high grid) in a setting where there is little by way of protective boundary-formation (low group), we have a rather strange form of social solidarity – fatalism: a form that is held together, not by the efforts of its members, but inadvertently, as it were, by some combination of the forces that are generated by other, much more active, forms of solidarity. Individuals who have no recourse to group membership, and at the same time find that their lives are largely determined by rules that are not of their making, are socially isolated (but not, we hasten to add, necessarily miserable). Here, policy outcomes emerge from an essentially uncontrollable process: the "garbage can" in its purest form.

These four forms of solidarity are not just "there", they are in endless contention with one another: like patches of differently coloured algae on the surface of a pond, each trying to extend itself by chewing bits off the others. Of course, as we move from one institutional setting to another, so the relative strengths of the solidarities, and their patterns of interaction, vary. Some institutions are strongly biased towards one solidarity, others to another and so on, but none is ever as organizationally pure as its organization would have you believe. In other words, subversion is inevitable, and forms of solidarity, whatever the institutional setting in which they are disporting themselves, always have to work hard to hold themselves together (the only exception being the fatalist solidarity which, as we have seen, is held together by outside forces). So it is never enough to be a member of a solidarity: you have to *know* you are a member of it (identity, in other words) and you have to *support* it, and this you do by entering into the norms and practices that you share will those whose identity you can recognize as being much the same as yours.

Individuals, of course, can find themselves members of different solidarities in different parts of their lives – different "transactional spheres" (workplace and home, for instance) – but they always have to engage in and support the solidarity they are part of, and this they do, among other things, by telling one another stories: stories that are very different from the stories that those in the rival solidarities are telling one another all the time. Once the members stop believing in their shared norms and practices, and once they stop telling one another the stories that unify them and hold them apart from others, the form of solidarity, by definition, ceases to exist. Organizational survival, therefore, depends on the continuous reproduction and legitimization of these various norms and practices, and if the variously organized patterns of social relations were not there, there simply would not *be* a policy process. So story-telling is not some regrettable aberration: without it, policy would not be possible.

We now begin to see how it is that each form of solidarity, if it is surviving, is providing its constituent individuals (who, in other parts of their lives, are likely the constitutents of other solidarities) with the appropriate set of cognitive tools – Allison's (1971) "perceptual lenses" – that allows them to perceive and understand how the world (both physical nature and human nature) is. These perceptual lenses then enable individuals to filter and select reality and experience, so that their interpretations of policy issues and solutions are congruent with the norms and practices of the form of social solidarity that, in that part of their lives, they are the supporters of.

Only the members of the fatalist solidarity are exempted from all this effort, and the overall result is a "triangular policy space" in which each apex is defined by the distinctive social construction of reality that both justifies and renders rational the behaviour of those who are the supporters of one of the three "active" solidarities and, at the same time, sets them apart from those who are similarly gathered around the other two apices[3].

What, the reader may well ask, does all this have to do with global climate change?

3. Three Stories About Global Climate Change

An analysis of global climate change policy debate in the mid-1990s revealed three policy stories. Each policy story provided a setting (in jargonese, the basic assumptions), villains (the policy problem), heroes (policy protagonists), and, of course, a moral (the policy solution). Depending on the socio-institutional context of the particular policy actor, each story emphasized different aspects of the climate change issue. What is more, each story defined itself in contradistinction to the other policy stories.

Profligacy

The first story begins by pointing to the profligate consumption and production patterns of the North as the fundamental cause of global climate change. Rich industrialized countries, so the argument goes, are recklessly pillaging the world's resources with little regard to the well-being of either the planet or the peoples of its poorer regions. Global climate change is more than an issue that is amenable to quick technical fixes; it is a fundamentally moral and ethical issue.

The setting for this story is a world in which everything is intricately connected with everything else. Whether this concerns human society or the natural world, this story

---

3 By this stage in our argument we have moved beyond the grid-group analytical scheme, that we used to set out the four solidarities, to the Cultural Theory with which that scheme is fully consistent. For an explicit outline of Cultural Theory see Thompson (this volume). The triangular policy space, and the various commitments by which it self-organizes, is depicted in Figure 3 of Thompson (this volume).

urges us to think of Planet Earth as a single living entity. Environmental degradation, then, is also an attack on human well-being. Humans, so the argument goes, have, until now, successfully deluded themselves that they can live apart from the natural environment. In reality, however, there is no place for humans outside nature and thus no particular reason for considering humans as superior to nature. In short, this story is set in an ecocentric world.

The villain, in the profligacy story, is the fundamentally inequitable structure of advanced industrial society. In particular, the profit motive and the obsession with economic growth – the driving forces of global capitalism – have not only brought us to the brink of ecological disaster; they have also distorted our understanding of both the natural and the social world. Global commerce and the advertising industry lead us to desire environmentally unsustainable products (bottled water, fast cars, or high calorie foods, for example) while our real human needs (living in harmony with nature and with each other) go unfulfilled. What is more, advanced capitalism distributes the spoils of global commerce highly inequitably. This is true within countries (the increasing gap between the rich classes and the poor classes) and between countries (the increasing gap between the affluent countries of the North and the destitute countries of the South). In short, prevailing structural inequalities have led to increasingly unsustainable patterns of consumption and production.

Since everything is connected to everything else, this story continues, we cannot properly understand environmental degradation unless we see it as a symptom of this wider social malaise. The way humans pollute, degrade, and destroy the natural world is merely a very visible indicator of the way they treat each other and particularly the weaker members of society. The logic that allows us to fell thousands of square kilometres of rainforest, to dump toxins in waterways, or pollute the air is precisely the same logic that produces racism, misogyny, and xenophobia. Tackling one problem inevitably implies tackling all the others.

The heroes of the profligacy story are those organizations and individuals who have managed to see through the chimera of progress in advanced industrial society. They are those groups and persons that understand that the fate of humans is inextricably linked to the fate of Planet Earth. The heroes understand that, in order to halt environmental degradation, we have to address the fundamental global inequities. In short, the heroes of the profligacy policy argument are those organizations of protest such as, most prominently, Greenpeace or Friends of the Earth: these organizations are strongly biased towards the egalitarian social solidarity outlined in the previous section.

What, then, is the moral of the profligacy story? Proponents of the profligacy story point to a number of solutions. In terms of immediate policy, the profligacy tale urges us to adopt the precautionary principle in all cases: unless policy actors can prove that a particular activity is innocuous to the environment, they should refrain from it. The underlying idea here is that the environment is precariously balanced on the brink of a precipice. The story further calls for drastic cuts in carbon dioxide emissions; since the industrialized North produces most CO_2 emissions, the onus is on advanced capitalist states to take action. Of course, this policy argument calls for a total and complete ban on chlorofluorocarbons.

Yet none of these measures, the story continues, is likely to be fruitful on its own. In order to really tackle the problem of global climate change we in the affluent North will have to fundamentally reform our political institutions and our unsustainable life-styles. Rather than professional old democracies and huge centralized administrations, the advocates of the profligacy story suggest we decentralize decision-making down to the grassroots level. Rather than continuing to produce ever-increasing amounts of waste, we should aim at conserving the fragile natural resources we have: we should, in a word, move from the idea of a waste society to the concept of a *conserving society*. Only then can we meet real human needs. What are real human needs? Simple, they are the needs of Planet Earth.

Prices

The second story locates the causes of global climate change in the relative prices of natural resources. Historically, prices have only poorly reflected the underlying economic scarcities; the result, plain for all to see, is a relative over-consumption of natural resources.

The setting of the prices and property rights tale is the world of markets and economic growth. Unlike the profligacy story, the prices diagnosis sees no reason to muddy the conceptual waters with extraneous considerations of social equality. Yes, it says, global climate change is an important issue, but it is an issue that is amenable to precise analytical treatment: it is, in short, a technical issue to which we can apply a technical discourse.

Economic growth, far from being a problem, is the sole source of salvation from environmental degradation. Environmental protection, the proponents of this policy argument contend, is a very costly business. In order, then, to be able to foot the huge bill for adjusting to a more sustainable economy, societies will have to command sufficient funds. These funds, in turn, will not materialize from thin air: only economic growth can provide the necessary resources to tackle the expensive task of greening the economy.

In sum, the prices tale takes places in a world determined by the invisible hand. Here, individuals know and can precisely rank their preferences. In the world of the prices and property rights story, individual pursuit of rational self-interest (economic utility) leads, as if by magic, to the optimal allocation of resources. If market forces are allowed to operate as they should, then resource prices will accurately reflect underlying scarcities: the price mechanism then keeps environment-degrading consumption in check. However, if someone (usually the misguided policy-maker) meddles with market forces, prices cannot reflect real scarcities; this gives rise to incentives for rational economic actors to over- or under-consume a particular resource.

The villain in the prices and property right story is misguided economic policy. Barriers to international trade, subsidies to inefficient national industries, as well as price and wage floors introduce distortions to the self-regulatory powers of the market. These distortions have historically led markets to place a monetary value on natural resources that belies the true market value. The result, the protagonists of this policy argument maintain, has been wholesale over-consumption and degradation of the natural world.

The heroes of the prices and property rights story are those institutions that understand the economics of resource consumption. In the global climate change debate, these institutions comprise players such as the World Bank or the OECD. In terms of the socio-institutional map outlined in the previous section, the heroes of this story are the market actors.

The moral of the prices story is as simple as its prognosis: in order to successfully face the challenge of global climate change, we have to "get the prices right". Unlike the profligacy story, the prices tale sees no necessity to restructure existing institutions. If it is the distortions of global, national and regional market mechanisms that undervalue natural resources then any climate change policy that fails to remove these distortions is "fundamentally flawed". Policy responses must work "with the market". Here, concrete policy proposals consist of both general measures, such as the liberalization of global trade, and more specific measures, such as carbon taxes or tradable emission permits.

Population

The third policy argument tells a story of uncontrolled population growth in the poorer regions of the world. Rapidly increasing population in the South, this story argues, is placing local and global eco-systems under pressures that are fast becoming dangerously uncontrollable: more people means more resource consumption, which inevitably leads to environmental degradation.

The setting of the population policy story differs slightly, but significantly, from the settings in the other two diagnoses. Like the protagonists of the profligacy story, the population policy argument maintains that global climate change is a moral issue. Human beings, due to their singular position in the natural world, are the custodians of Planet Earth; since civilization and technological progress has allowed us to understand the natural world more than other species, we have a moral obligation to apply this knowledge wisely. Unlike the profligacy story, the population tale assumes that humans have a special status outside natural processes. Similarly, the population story, like the proponents of the pricing argument, contends that human actions are rational. However, unlike the market argument, the population story tells us that the sum of individual rational actions can lead to irrational and detrimental outcomes. The population story, then, is set in a world that needs rational management in order to become sustainable. Yet, while the motive of rational management is an ethical duty to preserve the planet, the means of management are technical.

Economic growth, and the socio-economic system that underpins that growth, are necessary components in any global climate change policy response. However, economic growth in itself is no solution: it must be tempered, directed, and balanced by the judicious application of knowledge and judgement.

The villain in the population tale is uncontrolled population growth. Since each individual has a fixed set of basic human needs (such as food, shelter, security, etc.) and since these needs are then standardised at every level of socio-economic development, population increase, other things being equal, must lead to an increase in the aggregate demand for resources. Humans, the story insists, satisfy their basic human needs by consuming resources. It follows that population growth must lead to an increase in resource consumption: more people will produce more carbon dioxide in order to satisfy their basic needs. Given the limited nature of most resources, population growth must invariably lead to the over-consumption and degradation of natural resources.

The heroes of the population story are those institutions with both the organizational capacities (that is, the technical knowledge) and the "right" sense of moral responsibility for global climate change issues. In short, the global climate change issue should be left to "experts" situated in large-scale, well-organized administrations. In terms of the conceptual map of organizational types, the population story emerges from hierarchically structured institutions.

The moral of the population story is to rationally control population growth. In particular, this means the introduction of family planning and education in the countries most likely to suffer from rapid population growth. Here, the onus for action is quite clearly on the countries of the South. Rapid population growth has eroded societal management capacities; if we are to tackle the global climate change issue we must first establish the proper organizational preconditions.

4. Policy Implications

These three socio-institutional stories show that there are at least three different institutional versions of how climate change comes about, who is to blame, and what is to be done. Yet, what does this mean for climate change policy?

Irreducible Plurality

The three stories tell three plausible but conflicting tales of climate change. All three tales use reason and logic to argue their points. None of the tales is "wrong", in the sense of being implausible or incredible. Yet, at the same time, none of the stories is completely "right"; each argument focuses on those aspects of climate change for which there is a suitable solution cast within the terms of a particular form of organization.

Most importantly, these three policy discourses are not reducible to one another. No one of the policy arguments is a close substitute for the others. Nor are any of the stories' proponents ever likely to agree on the fundamental causes of, and solutions to, global climate change issue. And, since these implicitly transport a normative argument, namely that of the good life (either in markets, in enclaves, or in hierarchies), they are curiously immune to "enlightenment" by "scientific" facts: we cannot, in any scientific sense, "prove" or "falsify" policy stories. More pointedly, these stories define what sort of evidence counts as a legitimate fact and what type of knowledge is credible. The profligacy story discounts economic theory as the obfuscation of social inequalities and dismisses rational management as the reification of social relations. The tale of prices views holistic eco-centrism as amateur pop-science and pours scorn on the naïve belief in benign control. Last, the population story rejects laissez-faire economic theory as dangerously unrealistic and questions the scientific foundations of more holistic approaches.

This leaves us with a dynamic, plural and argumentative system of policy definition and policy-framing that policy-makers can ignore only at their cost, for two reasons. First, we have seen that each policy story thematizes a pertinent aspect of the climate change debate: very few, apart from extreme hard-liners (see below), would argue that Northern consumption habits, distorted prices, or population growth have no impact on global climate change at all. However, as we have seen, each story places a different emphasis on each aspect. Any global climate change policy, then, based on only one or two of these stories, will merely provide a response to a specific aspect of the global climate change problem. It will, in short, provide a partially effective response.

Second, and more significantly, each of the stories represents a political voice in the policy-process. Ignoring any of these voices means excluding them from policy-making. Within democratic polities, this inevitably leads to a loss of legitimacy. What is more, in democracies, dissenting voices will, eventually, force their way into the policy process. Neither the costs of acrimonious and vicious political conflict, nor the costs of suppressing dissenting voices, are particularly attractive: the former often leads to policy deadlock; the latter may well lead to a legitimacy crisis in the polity as a whole.

Three Implications for Global Climate Change Policy-Making

From these policy stories we can derive three interrelated implications for policy-makers.

1) Endemic Conflict: In a policy-process where politics matters (that is, in any policy process) there will always be at least three divergent but plausible stories that frame the issue, define the issue, and suggest solutions. Thus, conflict in policy-making processes is endemic, inevitable, and desirable, rather than pathological or deviant. Any policy process that does not take this into account does so at the risk of losing political legitimacy.
2) Plural Policy Responses: We have seen that each story tells a plausible, but selective, story. Any policy response modelled solely in terms of just one or two of these tales will be, at best, partial and, at worst, irrelevant.
3) Quality of Communication: Last, since policy-making is inherently conflictual, and since effective policy responses depend on the participation of all three

voices, policy outcomes crucially depend on the quality of the communication within the debate. A policy debate that can harness the inherent communicative and argumentative conflict between different story-tellers will profit most from the constructive interaction between different proponents. Conversely, a policy debate in which all three positions are sharply polarized will probably lead to policy-deadlock. This is a structural argument that concerns the implicit and explicit "rules" governing policy deliberation in a polity. If the "rules of the game" permit, or even force, policy actors to take different types of stories seriously, then what Sabatier and Jenkins-Smith call "policy-oriented learning" can take place. If this is not the case, then the policy debate will be an unconstructive "dialogue of the deaf".

5. Scenario Planning: One Tool from the Appropriate Policy Tool-Kit

The orthodoxy, in science-for-public-policy, is that you will get nowhere until you have established a clear definition of what the problem *is*. This tenet is then buttressed by another which stresses the importance of always maintaining a clear separation of facts and values. But this latter insistence requires that the unambiguous separation of facts and values is possible, and this is seldom the case. And the first tenet, since it inevitably gives credence to just one story – just one plausible account of how things are – and dismisses all the others, has, by our argument, to be reversed: if you are having to ask who's right (worse still, if you already *know* who's right) you are wrong!

The orthodox, optimisation-enabling, "simple metric" policy tools – cost:benefit analysis, for instance, general equilibrium modelling, probabilistic risk assessment and all those approaches that assume that uncertainty is merely the absence of certainty – eliminate plurality; right from the start, they get rid of that which, we have argued, must somehow be preserved. Scenario-planning, however, does precisely what we urge: it deliberately seeks out stories that are mutually irreducible, and it then aims to learn from all the incompatibilities that those stories give rise to. But, just because you are using scenarios, it does not follow that you are doing scenario *planning*.

- Shell, for instance, is famous for its pronouncing work on scenarios, yet it found itself debilitatingly surprised by Greenpeace's eleventh-hour intervention as it was about to dispose of the Brent Spar oil storage structure in the deep ocean. An analysis of Shell's published scenarios, however, has shown that they are all either market-based stories or hierarchical ones (Elkington and Trisoglio 1996). Just *one* egalitarian scenario would have alerted Shell to the risks it was blithely sailing into, but the absence of that solidarity at the company's decision-making levels meant that there were no "champions" for such preposterous stories.
- Ferenc Toth (personal communication) recounts how, at the first ever international meeting on "Surprise-Risk-Scenarios", in the mid-1980s, one brave soul came up with a storyline in which the Soviet Union collapsed and its satellites floated free. It was considered so patently absurd that it did not even get onto

the initial long list of stories to be considered. A few years later, of course, it happened!

- In the IPCC's case, things, until recently, have been even worse than they were in Shell. All the early scenarios (IPCC 1992), it has been shown, are based on the certitudes that underpin the hierarchical solidarity (Janssen 1996). Stories there are aplenty, but they are all the same story!

From these cautionary tales, one of which is rather uncomfortably close to home, we can see that plurality-preservation is something that is not easily achieved, and that it is emphatically not enough to say "Let's have some stories".

In other words, you are going to need some pretty robust – that is, theory-based – tools if you are to achieve the *requisite plurality*: the plurality that, we have argued, is adequately captured by our triangular policy space and its three storylines[4]. Here we will demonstrate that requisite plurality by setting out three scenarios of global climate change, each of which is the logical expansion of one of our three stories: prices, population and profligacy. There are, we should mention (since the approach we have taken is often dismissed for being critical of conventional policy approaches while not offering any constructive suggestions), several other plurality-preserving tools.[5]. So the tools *do* exist; the real challenge is to put down the much-relied-upon tools that, for all their obvious attractions, inevitably eliminate plurality.

The Practical Business of Hubris-Reduction

Scenarios, in *scenario planning* (and indeed in the everyday meaning of the word), are little stories: colourful, and often rather alarming, fleshings-out of different visions of the future. Different policy options can then be tried out against these scenarios, and those that are marvellously successful in one future but disastrous in others can be assessed against those that exhibit a certain robustness across them all. Scenario planning does not tell you which policy you should choose, but it does lead you, in a simple and instructive way, through the risks that exist within the *irreducible ignorance* (Ravetz 1993) that is so poorly handled by the conventional approach in which uncertainty is seen as nothing more than the absence of certainty[6].

---

4 The theory behind this triangular scheme, as we have already explained (end-note 3), is Cultural Theory: The Theory that explains the fourfold plurality that is so nicely captured by the grid-group analytical scheme. But, even without the theory, this grid-group scheme gives us a "heuristic device" – the triangular policy space – that, if employed, will enable the policy maker to avoid the sort of debilitating exclusions that are evident in our Brent Spar, our not-so-Surprise Rich and our IPCC examples.

5 A kit of eight tools for "policy in a complex and plurally perceived world" is set out in Thompson (1997).

6 Morgan and Henrian (1990) is the most thoroughly worked out example of this orthodoxy, but it has deep historical roots in economics: most notably Knight's (1921) conceptual framework – if you know what is going to happen that is *certainty*, if you do not know for certain but you know the odds that is *risk*, and if you do not even know the odds that is *uncertainty*.

Uncertainty, far from being an unproductive desert that will only bloom when we have managed to irrigate it with knowledge, is a resource: something that is all the time being colonized and fought over by contradictory certitudes: the shapes of the dose: response curve at low levels of ionizing radiation, for instance. Embedded in these seemingly technical assertions – that the curve bends away upwards (with lower doses of radiation being positively good for you), that it is the linear extrapolation of the established relationship, or that it bends away downwards (causing proportionately more harm at lower doses) – are profound and largely unquestioned assumptions about stability and change in nature: that Mother Nature can take anything we throw at her (the quadratic curve), that she is stable within limits, and therefore manageable once we understand those limits (the linear curve), and that she is so intricately interconnected that any intervention may provoke catastrophic collapse (the parabolic curve)[7].

Visions of the future are therefore *final causes*: attractors and repellers – way out there, ahead of us – that variously define in the here-and-now what is technically possible, socially desirable and morally acceptable. Science, of course, is essentially a programme for the systematic elimination of final causes, but these ones – the contending colonizations of uncertainty – cannot be eliminated (at least, not within the time-scale during which policies must be decided). And, if they cannot be eliminated, then it is unscientific to insist, as does conventional science-for-public-policy, that these final causes do not exist. Scenario planning, therefore, is best seen as a way of avoiding that sort of hubris.

The three scenarios presented here were developed, in 1994, in connection with the Battelle-initiated State-of-the-Art Report on Social Science and Global Climate Change (hereafter SOAR), the results of which have now been published as *Human Choice and Climate Change* (Rayner and Malone, 1998). It took six people (Alex Trisoglio, Rob Swart, Michael Thompson, Jan Rotmans, Hadi Dowlatabadi and Bert de Vries) just one day to produce them, and they are the first set of scenarios to be derived from an explicit theoretical base: Cultural Theory.

The basic idea is that the "contested terrain" of global climate change can be mapped by a 3 x 3 matrix. Each of the active solidarities – individualist, hierarchist and egalitarian – insists that the world is a way that the other two insist it is not, and each of them seeks to promote policies that are eminently sensible if that is how the world really is (but far from sensible – criminally irresponsible, even – if the world is one of the other ways) (Figure 1). Along the diagonal, therefore, we get three utopias, each of which is supported by a pair of *dystopias* (what would happen if the other solidarities were in charge). A full mapping, therefore, would require nine scenarios: three utopias and six dystopias, but the three we have generated are the utopias, (supported, here and there, with fragments from their associated dystopias).

---

7 For more on this particular set of colonizations of uncertainty see Thompson (1991) and Adams (1995, Chapter 3).

Figure 1:

		THE WORLD AS MANAGED BY		
		Individualist	Hierarchist	Egalitarian
THE WORLD	Individualist			
	Hierarchist			
BY	Egalitarian			

Each scenario begins with a short (one paragraph) outline of its key features and global climate change storyline, and then goes on to fill in the details under the 13 specific headings that, at that time, were going to be the chapters in the State-of-the-Art Report. In the event, the chapters turned out somewhat differently but these 13 headings are still discernible, still in the same order, in the first three volumes of *Human Choice and Climate Change*. It is hoped that these three scenarios will provide a useful framework: one that will encourage policymakers and policy analysts to incorporate the requisite plurality and, at the same time, enable them to identify the various biases that are built into existing scenarios (and models) that they wish to make use of.

SCENARIO 1: The Individualist's View of the World, and How to Manage It.

Name: Adaptation

Key words: Rapid change, technological innovation, no limits, cultural diversity, maximizing quality of life at individual level.

Problem definition: Global systems are resilient. The real challenge is overcoming obstacles to the innovation and growth that are key to improving quality of life.

Solution: Free markets to promote growth and innovation, focussing on quality of life at the individual level.

Summary: The global environmental concerns of the 1990s were clearly unfounded. With the planet resilient, humanity is free to concentrate on maximizing qualify of life. Free markets drive innovation. New technologies, especially in information, computing, biotechnology and cognitive science, revolutionize economic activity and lifestyles. A clean environment is pursued as part of a high quality of life, and enjoyed during free time. There is a wide diversity of lifestyles and kinds of work. At one end of the spectrum are low-consumption lifestyles, featuring preventative approaches to health, three days' work a week, and time for arts and leisure. At the other are high-income, hard-working professionals, with wealth to spend on exotic technologies and pursuits. Clean growth in the South leads to early demographic transitions, while economic growth and employment in the North is revitalized as entire economies begin to re-tool and rebuild with clean technologies.

Some considerations at the level of individual SOAR chapters:

1. Scientific Knowledge and Decision Making
   - Rapid innovation in all areas of science: research is increasingly sponsored by private companies, supported by governments for research with non-commercial applications.
   - Uncertainty is seen as an opportunity, not a threat.
   - Diversity in levels of decision making (delegation/subsidiarity).

2. Human Needs and Wants
   - Limitless quest for personal satisfaction and fulfilment in a world of abundance and diversity.
   - Reduced working hours for those who desire more time for hiking, arts, meditation, etc.
   - Others may work hard to afford a high-consumption lifestyle, including exotic goods such as private space craft, large-scale cosmetic restructuring, neural upgrading, etc.
   - Socializing in flexible networks.

3. Population, Health and Nutrition
   - Rate of population growth is declining because of the worldwide demographic transition brought about by economic growth.
   - People place growing importance on health as part of quality of life, either through prevention through individually chosen healthy lifestyles, or new bio-technological treatment.
   - Wide nutritional diversity, from organic produce to high-tech bioengineered foodstuffs for investment bankers too busy to eat.

4. Technology
   - New diverse technologies at appropriate scales (i.e. big or small to suit the task), notably communication, biotechnology, computing, and cognitive technologies to enhance brainpower (the cornerstone of value creation in the information economy).
   - Private modes of transport dominate, including remaining privatized public networks, but they are clean, and co-exist with bicycles and pedestrians.
   - High-tech communications facilities are ubiquitous.

5. Economic Activity
   - Worldwide free markets and trade, where the key government role is to ensure efficient market functioning, including strict anti-monopoly laws (ex-

tended, as foreshadowed in the Microsoft legal proceedings, so as to minimize technological "lock-in"[8].
- Growing diversity of working and employment patterns, with many taking on 3-day working weeks to allow greater focus on leisure, learning, etc., while others work 80 hour weeks.
- Information, skills, learning increasingly important.
- Equality of opportunity, maintenance of level playing field. But very different outcomes, with some choosing to work hard and become rich in material terms, others working less and becoming rich in community, spiritual, artistic, educational or health terms.
- Diversity of scale, form, sectors.
- New ways of measuring progress (towards improved quality of life) gradually replace the limited monetary and material bias of the system of the 1990s.
- Companies flexibly organized, avoiding large bureaucratic multinationals, which prove unable to innovate rapidly or keep costs low enough, or are broken up in anti-trust cases.

6. Energy and Industrial Systems
   - Diversity in supply, notably clean fossil technologies, but also renewables and new clean, inherently safe nuclear.
   - Environmental performance of technologies is key issue (as clean environment is a key component of quality of life). Environmental problems addressed by new technologies to remove pollution, increasing recycling, remanufacturing and closed-loop systems.

7. Land Use and Land Cover
   - Large increases in agricultural productivity, opening up large areas in North for nature development.
   - Productivity also increases in South, successfully addressing malnutrition.
   - Other economic values of forests are recognized, leading to the slowing of deforestation and increased regeneration. This is particularly true for tropical forests and their biodiversity which is central to a vigorous biotechnology industry.

8. Coastal Zones and Oceans
   - Coral reefs, wetlands, mangroves, fisheries etc., are preserved by markets because of their increasing scarcity and acknowledgement of their economic value (notably for tourism, leisure activity).
   - Demand for fish increasingly satisfied by fish farms, restocking the oceans if necessary.

9. Historical and Archaeological Analogues

---

8 This is a contemporary (2000) updating of this scenario.

- Reference to periods in history when (technological) development rapidly increased opportunity and improved standards of living. Point to collapses brought about by hierarchists or egalitarians interfering with the market.

10. Integrated Assessment Models
    - The role of models has changed with increasing computer power and usability. Models are now commodities that support individual decision making, personal planning and business, in the context of maximizing quality of life.
    - Commercial decision-making is increasingly controlled directly by intelligent computers.
    - Visualization, usability, accessibility important.

11. Institutions
    - Decreasing importance of nation states, their role focussing on enhancing equality of opportunity (for example, in access to education systems).
    - Shifting role of UN organizations towards facilitating market functioning (built on the World Trade Organization).
    - Increasing importance of function-based informal institutions, which are often transitory.

12. International Responses
    - If necessary, through alliances of private companies with the technological and managerial know-how.
    - Privatized security.
    - Militant nationalist/fundamentalist movements die out with economic development.

13. Natural Sciences
    - Increased emissions do lead to climate change, but the low climate sensitivity means that changes are gentle, slow, and non-destructive – not as bad as everyone thought. Ecosystems are resilient, agriculture benefits. Negative effects are addressed by adaptation.

SCENARIO 2: The Hierarchist's View of the World, and How to Manage It

Name: Wise Guidance
Key Words: Sustainability, managerialism, targets, steering, scientific expertise, international negotiation, control, optimization.
Problem Definition: Global environmental and economic systems are not under control.
Solution: Development of better control systems and institutional arrangements to manage the globe.
Summary: Governments play a major role in correcting social and environmental market externalities through regulation and economic instruments, thus steering the world onto a sustainable path that carefully balances economic and environmental objectives, and provides the best possi-

ble world for the greatest number of people. Environmental problems are serious but are resolved, without major social and economic upheavals, as countries muster the international commitments to identify the changes needed, and to implement those changes in a fair and enforceable way. Governments draw widely on the expertise and skills of industry, academia and the public in ensuring optimal decision-making, and the pursuit of cost-effective policies. The role of the United Nations and international government grows, to ensure effective co-ordination of economic and environmental policies. Careful stewardship enables continued but modest economic growth, needed to feed the growing population.

Some considerations at the level of individual SOAR chapters:

1. Scientific Knowledge and Decision Making
   - Science is more and more pooled in a centralized setting.
   - A great deal of scientific research is needed to reduce uncertainties, create solutions, and generate the information and careful weighing of costs and benefits needed for optimal decision-making.

2. Human Needs and Wants
   - Consumption styles and patterns are reshaped through education, regulation and economic incentives for environmentally-sound behaviour. Excesses are curtailed through strong legal sanctions.
   - There is a shift from rights to greater emphasis on responsibilities.
   - The working week is regulated to minimize unemployment through job sharing, public service, etc. Many government schemes are put in place to help people meet their needs.

3. Population, Health and Nutrition
   - Population growth is steered and controlled by education and family planning.
   - World population stabilizes at about 10-11 billion people.
   - The health system is state-provided, with high-tech and preventative measures. A lot of government money is spent on public goods, such as the health system and environmental services.
   - Better health habits are encouraged by public information and education. Consumption is changed by public service advertising campaigns, and by banning advertising that encourages unhealthy habits.

4. Technology
   - The transition to a cleaner economy is steered by technology forcing. Technologies include alternative clean sources of energy and zero-waste production processes.
   - Technological systems tend to be large-scale and centralized so as to optimise economies of scale.

- There is detailed technology assessment for new developments to ensure environmental compatibility, and a comprehensive system of permitting, labelling and regulatory control.
- Information and communications technologies are central to the control systems at all levels, with greater use of sophisticated computer technology to produce carefully regulated and optimal solutions, and also to ensure monitoring.
- There is an optimally structured mix between individual transport, and integrated and public transport systems.
- There is growing use of technology to control criminal activity and black markets, for example using sophisticated informatics.

5. Economic Activity
   - A comprehensive system of regulation and economic instruments is in place to internalize environmental costs, including large scale ecological tax reform. The revenues make up a substantial portion of tax income, and are redistributed through a number of environmental programmes and subsidies.
   - This gives rise to many trade distortions, which are unfortunate but unavoidable given the priority of attaining environmental and social goals, including employment, at the national level.
   - Large consortia of multinationals form to provide the necessary economic weight to compete effectively on world markets.
   - Large multinationals are central in most sectors, and form the centre of most people's lives: a community as well as a place of work. They provide a high degree of protection and quality of life, in exchange for the loyalty and service of employees. Corporations adopt voluntary "partnership" style environmental agreements with governments, allowing a smooth, controlled transition to sustainability.

6. Energy and Industrial Systems
   - Shift from fossil fuel-based systems to renewable and clean energy systems, ultimately in the form of hot fusion. Energy systems are centralized for optimal efficiency.
   - Energy use will increase, especially in the South. The increase will go hand in hand with cleaner forms of energy and pollution control.
   - Strategic government planning in close co-operation with industry, including R&D subsidies and incentives, succeeds in developing the key cleaner technologies for energy and industry.

7. Land-Use and Land Cover
   - Land-use management by planning: nested form of planning in the form of an internationally negotiated Land-Use Protocol.
   - Increasing productivity accompanied by designated preserves.
   - Massive and successful development co-operation programmes prevent deforestation, land degradation and soil erosion, and promote more efficient use of water. Managed top-down.

8. Coastal Zones and Oceans
   - Large-scale, highly engineered solutions.
   - Agreements and protocols for oceans, fresh water and enclosed coastal seas. Different protocols for different levels, but all consistent and integrated, including a highly developed quota system.

9. Historical and Archaeological Analogues
   - Institutional arrangements that tried to avert the tragedy of the commons.
   - Cumulative effect because of improved knowledge and learning by governments.

10. Integrated Assessment Models
    - Increasing role of integrated assessment models. These tend to be based on expert input, rather than the public, in order to optimise solutions to the complex economic and environmental problems.
    - Models used to develop indicators to be used in target-setting.
    - Integration of modelling and visualization.
    - Greater use of Geographical Information Systems and other computer tools for planning.

11. Institutions
    - International negotiations produce successful agreements for action on climate change, and United Nations Environment Programme (UNEP) is strengthened to form the World Environment Organization (WEO), which includes a structured network of international experts that work on a wide range of problems related to environment and development.
    - Growing importance and complexity of international organizations.
    - Formal structure is generic and recursive down to level of local government.

12. International Responses
    - By 2010, each country has more ambassadors to international organizations than to other nations.
    - More and more protocols established, improving co-ordination of environmental policies between countries.
    - Joint implementation.

13. Natural Sciences
    - Emissions continue to grow, but are gradually stabilised and eventually controlled through protocols under the Framework Convention on Climate Change. Negative impacts due to manifest climate change are compensated for by an internationally controlled aid programme. Climate change is a problem, but manageable. Growing knowledge and understanding help governments to plot the path that maximizes environmental returns for a given investment.

SCENARIO 3: The Egalitarian's View of the World, and How to Manage It

Name: Eco-Community

Key Words: Prevention, urgency, fragility, participation, new relationship with nature, decentralization, community, spirituality.

Problem Definition: Driven by government bureaucrats and reckless free markets that promote material consumption, society has overshot the world's natural limits, and also damaged community and the basis of society.

Solution: Rebuild community life, and drastically reduce human interventions in ecosystems, to meet spiritual and social needs.

Summary: Environmental problems are seen to be truly serious, necessitating radical change in direction, before it is too late. Growing dissatisfaction with, and distrust of, government and big business catalyses a new bottom-up approach to decision-making and action. The message of NGOs, grassroots initiatives and spiritual groups is seen to be right: humanity is an intrinsic part of the natural world, and must live in harmony with it. Societal organization focuses on communities and groups rather than on rigid hierarchical structures. Groups promote fairness and solidarity, forbidding the exploitation of fellow man or nature. This allows the regeneration of an active community that reduces the gaps between rich and poor, employed and unemployed, men and women. Rather than seeking wasteful and destructive consumption, people take a greater interest in their communities and in their inner lives, and realize that this is the source of true quality of life. Different communities evolve different responses. Some emphasise the use of sophisticated technologies and tools to minimize environmental impacts and improve social functioning. Others are more fundamentalist, rejecting technological "progress" with all the social and environmental damage it has brought, and adopting simpler and more holistic approaches to work, community, education, and the environment.

Some considerations at the level of individual SOAR chapters:

1. Scientific Knowledge and Decision Making
   - Decision-making emphasizes local participation and active local democracy, rather than centralized bureaucracies served by big science, which so often supported narrow commercial or militaristic aims in the past. Local experience and observation is vital, and there is a shift from "hard" science and technology to "softer" biological and social sciences.
   - Decision-making is much more community-based, with behaviour and impacts that are more visible, and hence more accountable. Governments still exist, but are truly democratic and participatory, and act to serve the needs of communities.
2. Human Needs and Wants
   - The priority is that the needs of all should be met, with as much equality as possible, ensuring a good quality of life across the whole community. Pov-

erty is ended, and the rapidly-growing gap between rich and poor of the 1990s has closed again. Redistributive taxation and social policies are dominant.

- Crime and insecurity are almost unknown, as people feel part of their communities again.
- Consumption is needs driven, hence much lower than in the 1990s. There is a spiritually motivated emphasis on non-material activities (work satisfaction, participation, personal growth, relationships, meditation). Conspicuous consumption, once seen as a sign of status, becomes socially unacceptable, much like smoking and fur coats in the 1990s.
- There is a focus on meeting needs, e.g. for mobility, comfort, joy, communication, knowledge, rather than on producing goods and services *per se*.
- Provision of needs is largely collectivized and localized, e.g. public transport and district heating.
- As the community grows in importance, people naturally think in terms of the needs of their neighbours, and voluntarily contribute to helping the needy or carrying out communally useful tasks, such as maintaining the built and natural environments, social work, etc.

3. Population, Health and Nutrition
   - Given the finite carrying capacity of the Earth, the world's population is slowing down and eventually decreasing. Responsible behaviour, recognising Mother Earth's carrying capacity and the rights of the other species who share her with us, ensures that by the middle of the 21st century the human population is around three or four billion. In the process, the basic needs of the poor are fulfilled by redressing the unequal distribution of resources.
   - Diets are largely vegetarian, with animal supplements depending on the local environment. In more technological communities, the supplement is a bio-techno-logical broth which uses trace minerals and sunlight.

4. Technology
   - Where technologies are prevalent, they emphasize subtle and sophisticated ways of working with nature, such as wind turbines based on tree design and lining, the use of medicine in harmony with bodily clocks or integrated pest management techniques.
   - They also focus on the spiritual rather than the material, with greater emphasis on traditional skills and insights, and craft work. There is greater emotional and spiritual depth, based on a sounder understanding of local ecosystems.
   - There is great use of e-mail and computer communications, to share solutions to technical, economic and other issues. Physical transport is minimized, with an international flight being a once-in-a-lifetime experience.
   - Rather than global uniformity, technologies are designed to fit closely with their surroundings, maximizing social and environmental harmony.

5. Economic Activity
   - Steady-state economies are essential. To stay within environmental limits, there are major changes in consumption. For example, by 2010 meat consumption per world citizen is 1oz per day, with 1 pint a day of milk, a reduction in Europe of 60-80%.
   - Sustainability of environmental systems and community solidarity take precedence over narrow economic considerations. Any proposed projects and activities have to demonstrate that they have no undesirable environmental or social impacts.
   - The economy is greatly dematerialized, with reduced materials flows. This is consistent with the growth of the information economy, where value is based on content rather than bulk.
   - Industrial trade is almost totally in high-tech products; food and building materials are largely derived from local resources. This is partly possible because of advances in using local biomass for fertilisers, plastics, etc. International commerce is partly trade-based, partly aid-based, on principles of solidarity and justice.
   - There is great emphasis on low resource intensity and self-sufficiency. In combination with lifestyle changes, resource flows drop by a factor of 10 to 100. High-tech applications such as solar cells and electronic devices are manufactured on a large scale.
   - Manufacturing and retail are run as community-based co-operatives, along the lines of Migros, Switzerland's largest supermarket chain in the 1990s. Since owners and consumers are the same people, there are incentives for low prices and good service, rather than high profits. The system also distributes resources much more fairly.

6. Energy and Industrial Systems
   - Communities have energy and industrial systems largely based on sustainable renewable resources. Eco-principles, like near-complete cycling of human-induced resource flows, dominate resource use and design, in areas from industry and transport to housing and household goods.
   - There is an emphasis on small scale, low capital-intensity solutions.
   - Changes in lifestyles and high-tech conservation reduce demand for energy, water and materials 5-10 fold compared to 1990s.
   - Bicycle-designed town planning is common, as is deep recycling based on separation by the consumer. Re-use, remanufacturing and highly durable, long lifetime products become the norm.
   - Extended community services, from libraries to social and sports facilities, create community while reducing the need for personal mass consumption.
   - Pollution of any sort is not tolerated.

7. Land-Use and Land Cover
   - Communities depend largely on local resources for food, with land management based on strict principles of sustainability. Some also produce food for export to nearby communities.

- Ecosystems are managed sustainably at the local level, some emphasizing sustain- able harvesting, others a return to wilderness. There are significant population migrations as people find they cannot live sustainably in many parts of the world that previously supported urban and industrial life. At the same time, those cities are made sustainable by city shrinking, pioneered in the Bay Area of Northern California in the 1990s.
- Undisturbed nature has a spiritual role as a teacher of life, as a way to reach out to the divine.

8. Coastal Zones and Oceans
   - Coral reefs, wetlands and mangroves preserved for what they are: an integral and valuable part of the biosphere. Economic considerations are not allowed to affect their [non] management.
   - Fish is a valuable food resource, all fisheries are managed sustainably. Forms of aquaculture are also used to provide highly productive protein sources.

9. Historical and Archaeological Analogues
   - Periods in which sustainable resource use principles dominated, because of low population density (as with hunter-gatherers) or spiritual and religious beliefs (some North American Indian tribes, sects like Quakers or certain ashrams in India).
   - Pioneer communities or networks which organise themselves around a shared vision: the Socratic and Pythagorean schools in ancient Greece, physicists in the quantum mechanics period, the Pugwash and Peace movements, the anarcho-communist movements (Walden, Kropotkin, etc.).

10. Integrated Assessment Models
    - The biosphere is too complex for a successful reductionist scientific attempt at “management“. A participatory research mode with awe and respect for the natural (and divine) world is dominant, with an emphasis on precautionary behaviour.
    - Models are used as tools to support and enrich real-world experience, e.g. in teaching and experimenting.

11. Institutions
    - Nation states become less important as the role of communities is strengthened, and children are brought up with the idea of one humanity.
    - People derive their community identity more from their river basin than from old-style national and regional boundaries.
    - E-mail and computer-based democracy ensure a highly transparent and democratic political system. Higher institutional levels are vestigial, and exist solely to support the community level.
    - All kinds of temporary coalitions and alliances are formed for regional issues, including environmental ones, evolving from present networks and NGOs. Although such arrangements may lack long-term stability, they prevent bureaucracy and ensure a vital and active involvement of communities in political affairs.

- Free rider behaviour is dealt with primarily through sanctions at the community level, or through community coalitions and alliances ("we don't do that kind of thing in our community").

12. Natural Sciences
    - Despite drastic emissions reductions, climate change materializes as expected because of apparent high climate sensitivity. Those who suffer are helped or compensated for by other communities on the basis of a transient inter-community Climate Change Victim Fund.

Contradiction, Viability and Change

In the course of generating these three scenarios, we all became aware of the contradictions that are built into each of them. This led to attempts to have scenarios rejected (or, at least, to have them returned for re-working). Cultural Theory, however, insists that no solidarity is viable on its own: each needs the others to do certain vital things for it that it is unable to do itself. And each, ultimately, needs the others to define itself against. Pointing out the contradictions at the heart of others' utopias, whilst (like Marx, with his strictures on capitalism) remaining blissfully oblivious to those at the heart of your own[9], is therefore all part of the never-ending process of contention that keeps the whole show on the road.

6. References

Adams, John, 1995: *Risk*. London: UCL Press.

Adler, Immanuell and Peter M. Haas, 1992: Conclusion: Epistemic Communities, World Order, and the Creation of a Reflective Research Program. *International Organisation*, 46, 1.

Allison, G.T., 1971: *Essence of Decision: Explaining the Cuban Missile Crisis*. Little, Brown and Co.

Clark, John and Aaron Wildvasky, 1990: *The Moral Collapse of Communism: Poland as a Cautionary Tale*. ICS Press, San Francisco, California.

Douglas, Mary, 1978: Cultural Bias. London Royal Anthropological Institute, Occasional Paper No. 35. Reprinted in Douglas (1982) *In the Active Voice*. London: Routledge and Kegan Paul.

Dryzek, J., 1990: *Discoursive Democracy*, Cambridge University Press.

Elkington, John and Alex Trisoglio, 1996: Developing Realistic Scenarios for the Environment: Lessons from Brent Spar. *Long Range Planning*, 29, 6: pp. 762-769.

Eriksen, E.O. and J. Wiegård, 1997: Conceptualising Politics: Strategic or Communicative Action? *Scandinavian Political Studies*, 20, 3: pp. 219-41.

Fischer, F. and J. Forester (eds.), 1993: *The Argumentative Turn in Policy Analysis and Planning*, UCL Press, London.

Habermas, J., 1987: *Theorie des Kommunikativen Handelns, Band 2* (Vierte Auflage), Frankfurt: Suhrkamp.

---

9 Clark and Wildavsky (1990), for instance, show how all Marx's predictions have turned out to be true, for communism!

Hood, C., 1998: *The Art of the State: Culture, Rhetoric and Public Management.* Clarendon Press, Oxford.

Janssen, M., 1996: *Meeting Targets: Tools to Support Integrated Assessment Modelling of Global Change.* Den Haag: CIP – Gegovens Koninklijke Bibliothek (ISBN 90-900 99 08-5).

Knight, F.H., 1921: *Risk, Uncertainty and Profit.* Republished 1965, Harper, New York.

Morgan, G.M. and M. Henrion, 1990: *Uncertainty: A Guide to Dealing with Uncertainty in Quantitative Risk and Policy Analysis*, Cambridge University Press.

Parsons, W., 1995: Public Policy: *An Introduction to the Theory and Practice of Policy Analysis*, Edward Elgar, Aldershot.

Ravetz, J., 1993: The Sin of Science: Ignorance of Ignorance. *Knowledge: Creation, Diffusion, Utilization*, 15, 2: pp. 157-165.

Rayner, S., and E.L. Malone (eds.) 1998: *Human Choice and Climate Change* (four volumes), Battelle Press, Columbus, Ohio.

Rayner, S., and M. Thompson 1998: Risk and Governance Part I: The Discourse of Climate Change. *Government and Opposition*, 33, 3: pp. 330-354.

Sabatier, P.A., 1987: Knowledge, Policy-Oriented Learning and Policy Change: An Advocacy Coalition Framework, *Knowledge: Creation, Diffusion, Utilization*, 8, 4: pp. 649-92.

Sabatier, P.A. and H.C. Jenkins-Smith (eds.), 1993: *Policy Change and Learning: An Advocacy Coalition Approach.* Westview, Boulder, Colorado.

Schwarz, M. and M. Thompson, 1985: Beyond the Politics of Interest. In: M. Grauer, M. Thompson, and A.P. Wierzbicki (eds.): *Plural Rationality and Interactive Decision Processes*, Lecture Notes in Economics and Mathematical Systems 248, Springer-Verlag, Berlin, 22-36.

Smith, Martin J., 1993: *Pressure, Power and Policy: State Autonomy and Policy Networks in Britain and United States*, Harvester-Wheatsheaf, London.

Stone, Deborah, 1997: *Policy Paradox: The Art of Political Decision Making*, W.W. Norton and Company, London.

Thompson, M., 1991: Plural Rationalities: The Rudiments of a Practical Science of the Inchoate. In J.A. Hansen (ed.) *Environmental Concerns: An Interdisciplinary Exercise*, Elsevier Applied Science, London, pp. 243-256.

Thompson, M., 1997: Cultural Theory and Integrated Assessment. *Environmental Modeling and Assessment*, 2: pp. 139-150.

Thompson, M. and S. Ney, 1999: Consulting the Frogs: The Normative Implications of Cultural Theory. In: M. Thompson, G. Grendstad and P. Selle (eds.) *Cultural Theory As Political Science*, Routledge, London, pp. 206-223.

Consumption, Motivation And Choice Across Scale: Consequences For Selected Target Groups

Michael Thompson

That culture matters, and that culture is not adequately taken account of in economic reasoning, is nicely illustrated by Tariq Banuri's famous wager: that the price of pork could go to zero, or even become negative, and nobody in Pakistan would start eating it.[1] Behaviour – of all sorts, not just consumption behaviour – seems often to be fixed in ways that the economist cannot explain. This realization leads us to all sorts of "what ifs" – unlikely perhaps, but who can tell – that if they came to pass, either by design or by accident, would have tremendous implications for climate change, and for our efforts to do something about it.

- What if China took to the motor car to the extent that the developed world already has?
- What if Russia went vegetarian?
- What if Indians started to eat their cows?
- What if "down-shifting" really caught on in the United States?
- What if there was a sudden, highly cost-effective break-through in renewable energy technology: cold fusion, say, or ocean geo-thermal?[2]

I could go on, but my point is simply that culture *does* matter, and that we need, somehow, to understand how and why it matters.

1. The Rules Of The Cultural Method

Man, we all know, is a cultural animal, yet most theories in social science are (or, rather, appear to be) non-cultural. For instance, all that distinguishes man from other animals, according to economists, is his insatiability. Give a dog a bone, or a cow its cud to chew, and it will be happy, but a human, these theorists insist, will always prefer a larger bundle of goods to a smaller one: a stance that is nicely satirized in

---

1 This wager, and its significance for theories of culture, is explored in the introductory chapter (pp.1-18) of Ellis and Thompson (1997).

2 The cultural nature of this "What if?" may be less immediately obvious than that of the others. But technological evolution - which possible paths end up being taken - is massively influenced by the convictions as to what is physically possible, socially desirable and morally justifiable that accompany each of the solidarities, and by the resultant *technological commitments* - to large-scale/capital-intensive (hierarchy), to small-scale/labour-intensive (egalitarianism) and so on - and by institutional arrangements that exclude one or more of these "voices" from the decision making process. (See Schwarz and Thompson, 1990; Tranvik, Thompson and Selle, 1999.)

E. Jochem et al. (eds.), Society, Behaviour, and Climate Change Mitigation, 93–108.

the apocryphal exclamation "When I hear the word 'culture' I reach for my utility function!".

Though we may smile at the economist's astounding simplification of man's cultural nature, the *non-satiety requirement* (as it is called, and it is absolutely central to economic theory)[3] does take him quite a long way – his model of economizing, in which needs are always outstripping resources, *does* fit quite a lot of human behaviour.[4] And he is certainly to be congratulated for his resistance to most attempts at taking some sort of explicit account of culture: a resistance that can be summarized in my own preferred misquotation of Herman Goering's immortal words:[5]

When I hear the word "culture" invoked as:

- An uncaused cause
- An explanation of last resort
- A veto on comparison,

I reach for my revolver.

In fact, those of us who do seek to provide cultural explanations, and therefore are convinced that culture *does* matter, see these three stipulations as "the rules of the cultural method" (Thompson, Ellis and Wildavsky, 1992). Let me explain a little more about them, because they are rules that are not easily obeyed and that, moreover, are currently much disregarded.

- **Culture as an uncaused cause**. These are explanations of the form: "Why did he do that?"; "Because his culture told him to". The invocation of "Asian values", or statements such as "Japan is a high-trust society; the United States a low-trust society" or that "the Judeo-Christian tradition is anthropocentric and can only justify environmental protection as resource management", are examples of this solecism. Though often dressed up in impressive swathes of reason-

3 Economists are careful not to deny the truism that you can have too much of a good thing. People, they argue, can become pig-sick of smoked salmon, say, and they are careful to say that the insatiability applies to things in the plural, not to this or that particular thing. Hence the "bundle of goods" form of words (see Awh 1976). We should also note that if people were not insatiable they would not have to prioritize their ever-proliferating needs so as to bring them within their more limited resources, and if they didn't do that the whole edifice of neo-classical economics would collapse. So it is the theory that *requires* the non-satiety.

4 I say "he" because it usually is, but the writer Ayn Rand, who had "hes" such as Milton Friedman and Alan Greenspan as disciples, would be a worthy exception.

5 Of course, it was "high culture" - the arts, literature and so on - rather than the anthropological notion (all the things that we have that monkeys haven't) that Goering was talking about. On top of that there is no record of him ever saying it! It was a character in a 1930s play - "Schlageter", by Hans Johst - who said it, though I am assured that the character in question did bear an uncanny resemblance to Hermann Goering. Also, what he actually said, when translated, was: "When I hear the word 'culture' I release the safety catch on my Browning" [the famous brand of revolver, that is, not the British poet].

ing, these simply are not explanations: just elaborate ways of saying "I don't know".

- **Culture as an explanation of last resort**. This is when culture is dragged in only when other explanations – economic, demographic, ecological, organizational, political and so on – are inadequate. Non-cultural explanations, of course, are often advanced in relation to environmental matters; indeed they dominate: the PRED framing (Population, Resources, Environment and Development), for instance, the "IPAT equation" (environmental Impact equals some multiplication of Population, Affluence and Technology [Ehrlich and Holdren, 1974]) and pretty well all the computer-based models that are so relied on in environmental policy making (and that swallow up so much of the available funding). Such approaches, since they take no account of cognition – seeing *and* knowing – are hopelessly reductionist, and treat people as essentially no different from cattle. They could never, for instance, account for what happened in Greenland during the last mini-ice-age, when the Inuit adapted and prospered and the Vikings stuck to their livestock-rearing and died out.
- **Culture as a veto on comparison**. The idea here is that each culture (and each sub-culture) is unique and can only be understood in its own terms. This idea goes back to Wittgenstein's "language games" and is now most firmly entrenched in *interpretive sociology* - most famously in Clifford Geertz's notion of "thick description". But, as Harry Eckstein (1997, 27), one of the contributors to the book *Culture Matters*, observes, thick descriptions, in the absence of any attempts to test and compare, are just "very high-level travel literature".

So, with non-cultural approaches (which currently dominate) ruled out as non-starters, and with most of the current cultural approaches being rejected as well by one or more of these three rules of the cultural method, what are we left with? That is the question I will try to answer in this paper.

My answer will be framed in terms of what is called Cultural Theory (more properly the "theory of socio-cultural viability" [Thompson, Ellis and Wildavsky, 1990]) but I should first mention that, in the book *Culture Matters* (1997) that Richard Ellis and I edited, less than half the contributors are Cultural Theorists, even though they all subscribe to the rules of the cultural method. So there are quite a few other ways of insisting that culture matters, and of obeying these rules, as well as the one I will be outlining. In other words, if you like the line of argument, but not the theory by which it is framed, that is not too serious a problem. On the other hand, if you don't like either then we *do* have a serious problem!

2. How, then, *does* culture matter?

We can avoid these three pitfalls - culture as an uncaused cause, as an explanation of last resort, and as a veto on comparison - in the following way.

- Beliefs and values do not just float around, with people choosing a bit of this and a bit of that. They are closely tied to distinctive *patterns of social relations* and to the distinctive *ways of behaving* that those beliefs and values justify. Cultural Theorists refer to each of these mutually supportive comings-together of *cultural biases*, patterns of *social relations* and *behavioural strategies* as a "form of social solidarity".

- Beliefs and values, therefore, are not just an explanatory "add-on"; they are essential components of economic, ecological, demographic, organizational, political and so on explanations.
- We *can* distinguish similarities and differences across cultures in terms of a small number of universally valid forms of social solidarity.

Well, as I'm sure most of you know, these forms of social solidarity can be set out by completing the typology that is implicit in the conventional social science distinction between *hierarchies* and *markets*. Markets institute equality (of opportunity) and promote competition, whilst hierarchies institute inequality (status differences, such as those between "experts" and "lay people")[6] and set all sorts of limits on competition, and this means that there are two other permutations: equality without competition (which we call *egalitarianism*) and inequality with competition (which we call *fatalism*). And these four solidarities, as I will explain in a moment, entail a fifth and rather unengaged solidarity - *autonomy* - that is characterized by the hermit.

Rather than drag you all the way through Cultural Theory, let me try and bring this typology alive by taking a quick look at some of the subtle interactions of forms of social solidarity, ways of knowing and family-level strategies that characterize the lives of Himalayan villagers: one of the "selected target groups" in the title I have been given.

This target group, I should explain, is particularly relevant for climate change because it has been wrongly blamed for rampant deforestation which, of course, if it is happening, is a major source of increased atmospheric carbon dioxide. A correct understanding of what these villagers are doing with their environment is therefore a prerequisite for an adequate definition of this part of the CO_2 problem, and for the design of effective solutions to it.[7]

---

6 I mention this because we are billed as an "expert meeting", which suggests that the IPCC's cultural bias is very much towards the hierarchical solidarity.

7 Briefly, the orthodox view, for more than twenty years now, has been framed in terms of an enormous "vicious circle": a mushrooming population is being forced to clear the forest (which already survives only on the steeper slopes) to create more agricultural land to feed itself. In doing this, it inevitably increases erosion and lessens the fertility of those fields that have not yet disappeared into the mountain torrents (because the fuelwood shortage leads to the farmers burning the animal dung that, in less stressed times, they used as manure) thereby forcing them to clear even more forest from even steeper and even more erosion-prone slopes. On and on.

The model of the person that is built into this definition of the problem is the "ignorant and fecund peasant": a model that, it has now been shown (see, for instance, Thompson, Warburton and Hatley 1986; Ives and Messerli 1989; Thompson 1998b), does not fit at all well with the actual behaviour of Nepali farmers (who, for instance, are quickly regenerating their now denationalized village forests and, in the meantime, have bridged the fuelwood and fodder gap by allowing trees to grow on the banks between their terraced, and privately-owned, rice-fields). Indeed, the lack of fit is so pronounced that the problem definition, and the solutions - the policies - it has given rise to , have now been completely discredited. It is, it is now increasingly recognized, the institutional capabilities of the people who live in the Himalaya, not their numbers, that determine whether the crucial spirals are

- When, as often happens, these subsistence farmers set about making themselves a little more secure by engaging in a spot of "off-farm employment" - portering for tourist treks, or going off on trading expeditions over into Tibet or down into India - they are binding themselves into one well-known form of solidarity: the market, but Cultural Theorists (in order to emphasize its cultural and institutional nature) label it *individualism* (characterized by symmetrical transactions and an absence of accountability, and supported by the convictions that nature can take anything we throw at it and that man is irredeemably self-seeking: Figure 1).
- But when the villagers appoint their forest guardians (*naua*), and respect their judicious stipulations about this year's level of permissible use of the village forest, they are binding themselves into a rather small-scale and face-to-face *hierarchy* (characterized by asymmetrical transactions and accountability, and supported by the convictions that nature is unstable when pushed beyond discoverable limits and that man, though deeply flawed,[8] can be redeemed by firm and nurturing institutions: Figure 1).
- And when, as happens with the famous Chipko Movement, they drop everything and rush out to hug the trees (*chipko* means "to stick") to prevent them from being appropriated by rapacious logging contractors, aided and abetted by corrupt forestry officers, they are binding themselves into an *egalitarian* form of solidarity (characterized by symmetrical transactions and accountability, and supported by the convictions that nature is so fragile there are no safe limits and that man is inherently caring and sharing [until corrupted by those nasty, coercive and inequality-creating institutions: markets and hierarchies]: Figure 1).
- Finally, when these resourceful and social capital-rich patterns of interaction fail (as one of them – the hierarchical arrangements for managing the village forests – did, following the nationalization of Nepal's forests in the 1950s) then the distinctive morality that underpins each of them is destroyed, the controls that are built into each of them break down, trust ebbs away, and purposive actions give way to *fatalism* (characterized by asymmetrical transactions and the absence of accountability, and supported by the convictions that nature operates without rhyme or reason and that man is everywhere fickle and untrustworthy: Figure 1).
- Nor, in the Himalaya, is even that the end of it! All four solidarities, despite their impressive differences, are fuelled by desire and, therefore, are mired in ignorance: something that is constantly being pointed out to our culturally pluralized villager by the Buddhist *lamas* and Hindu *sadhus* who are to be found on every mountain-top and at every source of the Ganges. When, as quite often happens, our farmer, recently widowed or widowered, hands over the house and land to the youngest son and withdraws to a cave in some nearby cliff-face he or she

upwards or downwards. From being the root-cause of the problem, the Himalayan farmer now becomes a key component in the solution. The farmers, of course, want the trees for their vital role within their Himalayan farming systems, but if they are there for that local security-enhancing purpose they (and their sequestered carbon) are also there for the global security-enhancing purpose of mitigating climate change.

8 "Deeply flawed", jumping from the Himalaya to the hierarchical reaches of British social life, were the Archbishop of Canterbury's very words following the death, in a car accident, of Diana, Princess of Wales.

disengages (ideally, for seven years, seven months and seven days) from all four "coercive" solidarities and binds himself or herself into *autonomy* (characterized

Figure 1: The Five Forms of Solidarity: the 5 ways of organizing and the 5 social constructions (myths of physical and human nature) by which they are sustained.

ASYMMETRICAL TRANSACTIONS

FATALISM	HIERARCHY
Physical Nature Capricious: operates without rhyme or reason. **Human Nature** Fickle: people are inherently untrustworthy and there is no fairness in this life.	**Physical Nature** Perverse/ Tolerant: stable and predictable within discoverable limits; unstable beyond those limits. **Human Nature** Malleable: deeply flawed but redeemable by firm, nurturing and long-lasting institutions that can be trusted and that distribute fairly (i.e. by rank and station).

AUTONOMY

UNFETTERED COMPETITION (no accountability)	The Physical/Human nature distinction is false; both are subsumed within an endless cycle that, in turn, subsumes the other 4 solidarities' myths of physical and human nature.	FETTERED COMPETITION (accountability)

INDIVIDUALISM	EGALITARIANISM
Physical Nature Benign: can be counted on to bounce back. **Human Nature** Self-seeking: "Hidden hand" ensures that people prosper only when others benefit as well. Trust others till they give you reason not to. Only fair that those who put most in get most out.	**Physical Nature** Ephemeral: least jolt may trigger total collapse: "tread lightly on the earth". **Human Nature** Caring and sharing. Corrupting institutions (markets and hierarchies) to be distrusted. Only equality of result (not of opportunity) is fair.

SYMMETRICAL TRANSACTIONS

by the minimization of all transactions and the internalization of accountability, and supported by the not-easily-communicated conviction that all dualisms – of which the distinction between physical and human nature is one – are false: Figure 1).[9]

So here, in essential outline (and sketched with the help of that impressive *dividual*, the Himalayan villager) is the Cultural Theory typology: a typology that provides valuable guidance through a complex and not easily mapped terrain: the interplay, on the one hand, of micro and macro (Himalayan household and Nepalese state, for instance) and, on the other hand, of us humans and the rest of creation.

Security, we begin to see, has to do with the interaction of these different solidarity-induced styles, *at* each and every scale level – household to nation state to international regime – and *between* all those levels (Figure 2).

Figure 2: Cultural Theory's multi-layered template: five cultural styles at every social scale level.

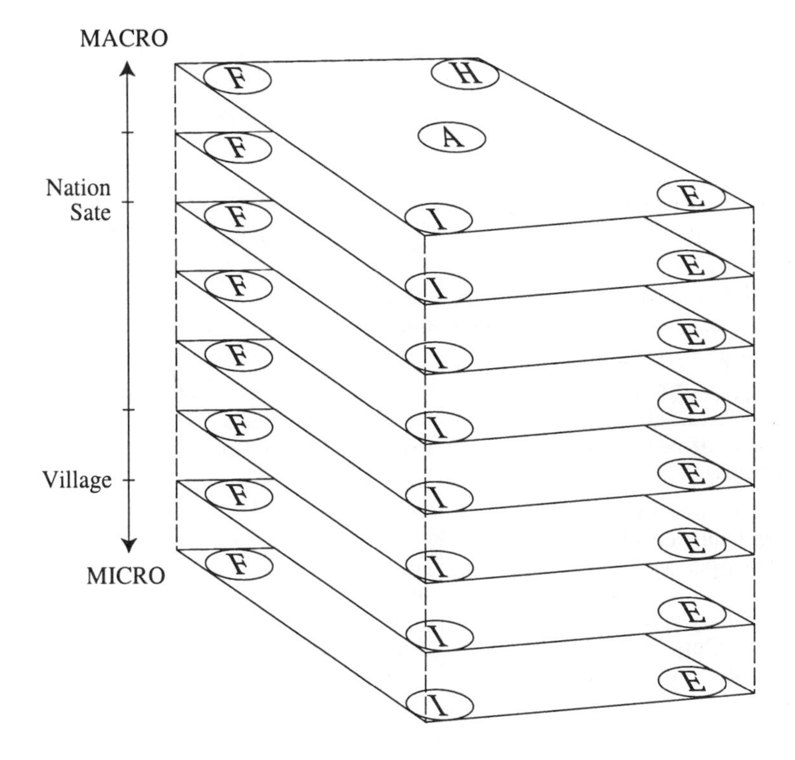

9 Hence the emphasis, in this solidarity, on meditation. The famous Tibetan hermit, Milerepa, kept his distance from a visiting, and hierarchical, academic by exclaiming: "Accustomed long to meditate upon the Wispered Chosen Truths, knowledge of erring ignorance I've lost", which, while low on communicative content, had the desired effect (to get rid of the unwelcome visitor) by making clear that the social constructions that sustained his autonomy *were* incommunicable. For some explana-

Things, in consequence, in any particular transactional instance, can go swimmingly in just one way - right scale and right style – and they can go into security-sapping reverse in three different ways – right scale/wrong style, right style/wrong scale and wrong scale/wrong style - and the Himalaya provide us with many excellent examples of all these (see Thompson 1998b). The collapse of Nepal's village forests after their 1950s nationalization, for instance, is a clear case of right style (hierarchy) but wrong scale (state when it should have been village).[10]
Beyond the Himalaya, these security-sapping reverses are, if anything, even more dramatic. The World Trade Organization, for instance, as it is currently constituted and promoted, is insisting (implicitly) that all transactions be in the individualist style and, moreover, (explicitly) that all these market-based transactions be at the global scale.[11] In other words, it is insisting that just one of the umpteen slots in Figure 2 (the individualist style on the topmost layer) is right and all the others are wrong, everywhere and always. That, Cultural Theory suggests, *cannot* be true. Indeed, there is now some compelling empirical evidence from another mountainous country - Austria - of its wrongness in relation to the handling of hazardous waste. Questionnaire-based research indicates that those market transactions that *are* appropriate for hazardous waste should be restricted to the national level, there being an overwhelming conviction among Austrian citizens that the exporting of hazardous waste to other countries is profoundly unfair (Linnerooth-Bayer 1999; Thompson 1998a).

Security enhancement which, I am suggesting, is what we need to focus on if we are to do anything constructive about climate change, therefore has to do with all those modish commitments – reflexivity, deliberation, multi-loop learning, clumsy institutions, robust policy design, the argumentative style of democracy, and so on[12] – that will be needed if we are to increase the number of occasions on which we get both the style and the scale right and decrease the number of occasions on which we get one, the other or both wrong. That, in a rather small nutshell – a nutshell that has been arrived at by a rather swift excursion to the southern slopes of the Himalaya - is the novel social science perspective that is opened up once we start taking serious and valid account of culture.

The above paragraph, I now realize, is my conclusion, which rather puts me in the company of that French director (was it Robbe-Grillet or Luc-Goddard?) who said

tion of how this seemingly asocial solidarity enters into and shapes Himalayan society see Thompson (1982).

10 But the state is now doing the appropriate thing. The forestry service is now helping the villagers to repossess their now de-nationalised forests while, at the same time, being careful to leave the reinvention of the commons-managing institutions to the villagers. The forestry officers offer help only when it is requested (subsidiarity) and concentrate on developing techniques – such as the regeneration of forests on those denuded slopes that have lost all their soil – that are not yet within the competence of the villagers (see Thompson 1998b).

11 This was written, in November 1999, just before the "Battle of Seattle" which, of course, has left the WTO *un*constituted and *un*promoted.

12 All of which, we should observe, were, prior to the Battle of Seattle, missing from the World Trade Organization, and none of which is too visible in the IPCC, as currently constituted.

that "a film should have a beginning, a middle and an end, but not necessarily in that order". So let me now finish with my middle: a number of examples – each dealing with a "selected target group" – of this cultural method in action.

3. Some Applications Of The Cultural Method

Cultural Theory, in its strongest form, argues that these five forms of solidarity are all the forms there are (the *impossibility theorem*, as it is called) and, moreover, that each is viable only in an environment that contains the others (the *requisite variety condition*, as it is called). It therefore predicts that one will find these five solidarities, in varying strengths and patterns of interaction, in *any* society and at *any* scale level: household, firm, political party, nation state, international regime or whatever (as, of course, is depicted in Figure 2). For example, if one takes the largest scale of all, one will find that the debate over global climate change has an irreducibly triangular structure (triangular rather than five-sided because the fatalist solidarity has no voice and the autonomous solidarity elects not to speak; only the three "active" solidarities are involved) (Figure 3).

Figure 3: The Contested Terrain of Climate Change Policy

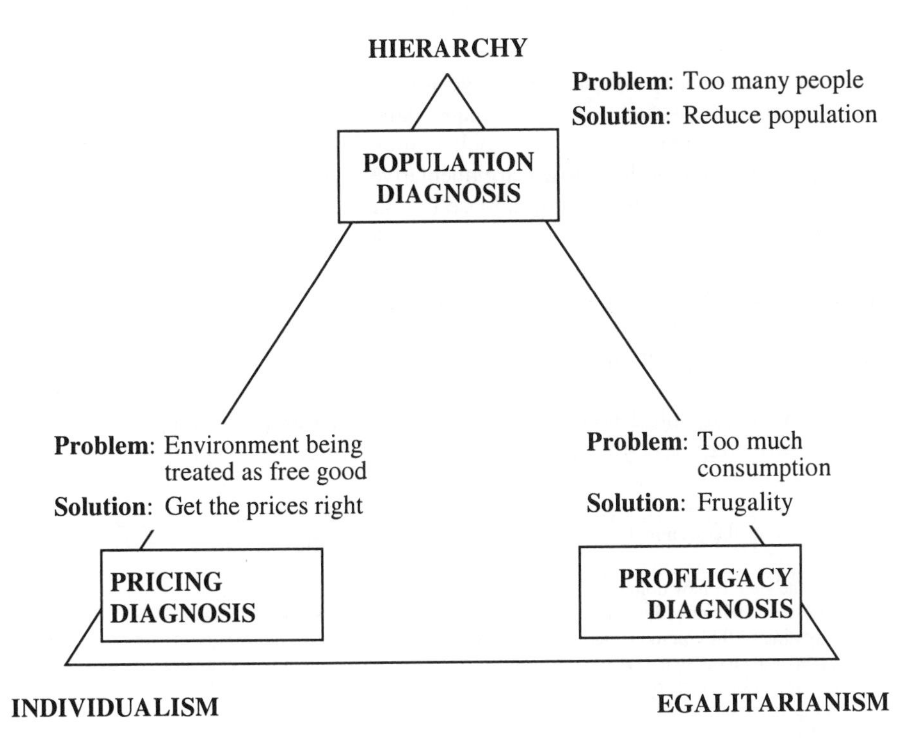

This, as I'm sure you know, is the key diagram from the 4-volume study *Human Choice and Climate Change* (Rayner and Malone, 1998). "Key" because it is from this diagram that both the social science framing and the policy implications are derived. However, I will not go into all that here (it being already set out at length in *Human Choice and Climate Change*, particularly in Volume 1, Chapter 4: "Cultural Discourses") but will press on with a couple more examples of the cultural method that are at much lower scale levels: consumption styles in Britain and India.

Household Consumption Styles in Britain

Cultural Theory predicts that each solidarity will give rise to (and, in its turn, will be supported by) a distinctive consumption style: a style that then translates directly into a shopping basket the contents of which are markedly different from those that will be found in the hands of those who have bound themselves into the other solidarities.

- the *cosmopolitan* style of the individualist (for whom the world is his oyster, and a thing of beauty a joy for a fortnight). This, of course, is the consumption style that fits the economist's non-satiety requirement.
- the *traditionalist* style of the hierarchist (who anchors his stratified collectivity in the weight of history)
- the *naturalist* style of the egalitarian (who, in rejecting artifice and excess, seeks to bring human demands down within nature's frugal limits)
- the *isolated* style of the fatalist (for whom nature operates without rhyme or reason, suggesting therefore that there is no point in trying to manage needs and resources in any way).

A study of 220 British households (Dake and Thompson, 1993; 1999) supports this hypothesis, and its results are summarized in Figure 4.

Depicted here are the results from two very different methodologies. The letters in the square boxes (I, F, H and E) mark the "centres of gravity" of the households (77 in all, randomly selected from the 220) that were identified, by *anthropological interview*, as being individualist, fatalist, hierarchist and egalitarian.[13] The numbered dots that form the star-like arrangement are the same households' responses to the *questionnaire* that was "double-blinded" with the interviews. In other words, the interviewers (myself and either Kathy Guy or Sylvia Lancelotte) did not know how the households had come out in the questionnaire, and the person who analysed the questionnaire results (Karl Dake) did not know how the households had come out in the interviews. I mention this, first, because such double-blinding provides rather a stern test of the hypothesis and, second, because the gulf between the "participant observation" that is inherent in the interviews and the "scientific detachment" that characterizes the questionnaire usually ensures that just one of these methodologies is chosen and the other rejected. Not the least of Cultural Theory's charms is that it encourages us not to rush off and join one side or the other (though Karl Dake and I have been virtually "drummed out" of our disciplines – quantitative psychology and

---

[13] The square-boxed A is the centroid of those households that were judged to be autonomous in the anthropological interviews. I will ignore these here because we have not yet developed the questionnaire to the point where it can tap in to the autonomous consumption style.

social anthropology, respectively – for not making that choice). I should also point out that the two dimensions in this picture are not the same as those that organize the basic Cultural Theory diagram (Figure 1); they are merely an artefact of the technique (discriminant function analysis) that has been used to reveal the pattern in the data.

Figure 4: Household Consumption Styles in Britain

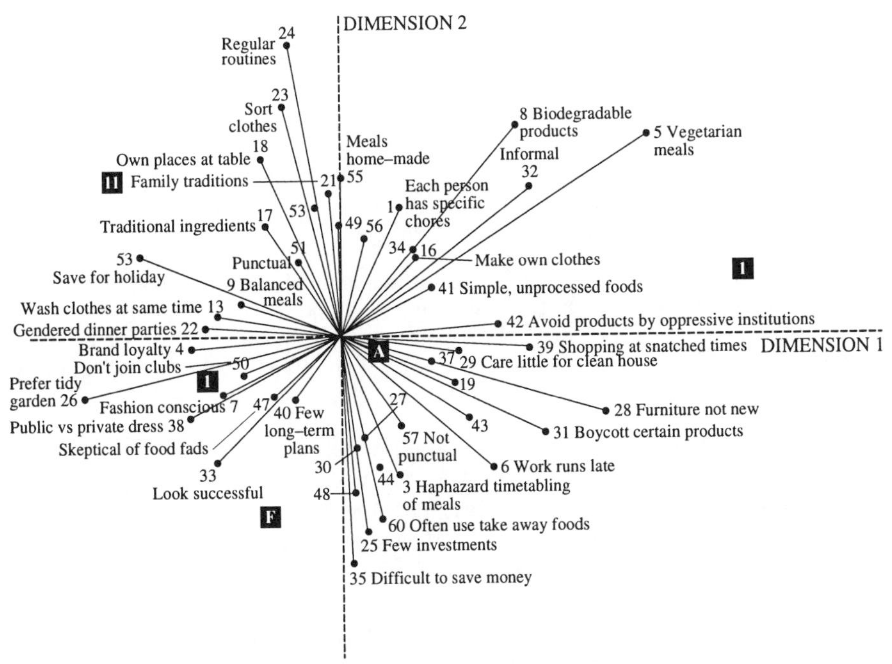

Note: n = 77 residents of London, Lancashire and Merseyside, England. Two discriminant functions show responses to 60 survey items describing household behaviour as well as the centroids for hierarchical (H), individualist (I), Egalitarian (E), fatalist (F), and autonomous (A) households as classified by anthropological interview. The horizontal dimension accounts for 64.8 % of the variance of self-reported behaviour; the vertical dimension accounts for an additional 17.7 %. The interviews and the surveys were conducted double-blind. These findings therefore suggest a high degree of convergence for these two independent sources of information. Some points and some labels have been omitted for clarity.

To read off the different consumption styles that accompany and support each of the four solidarities, you should focus on those rays that lie roughly in the direction of the appropriate "centre of gravity" *and* on those rays that lie in the diametrically opposite direction.

- *Individualists* are fashion conscious, like to look successful, prefer a tidy garden and don't join clubs (in the manner of Groucho Marx who wouldn't join any club that would have him as a member). They do not go in for vegetarianism, biodegradable products or informality, nor do they allot specific chores within the household or go out of their way to avoid fascist vegetables and the like – the products, as we coyly put it (we were doing this work for the Anglo-Dutch multinational, Unilever), of "oppressive institutions".

- *Hierarchists* gender their dinner parties (different wines for the ladies and the gentlemen, for instance). They use traditional ingredients, have their own places at table, are sticklers for punctuality and wash their clothes on "wash-day", having first sorted them out according to colour and fabric. They do not boycott certain products, nor do they find that work often runs late, nor do they go in for take-away meals. They are not prepared to put up with old furniture, nor are they comfortable with a house that is untidy.
- *Egalitarians* are about as opposed to both individualists and hierarchists as it is possible for them to get without disappearing off the diagram. Communality, together with unprocessed foods and biodegradable products (which, moreover, should be made by institutions that are not oppressive), is what they look for: an informal, joining, vegetarian sort of a life, in a pleasantly scruffy house filled with furniture that is not new and surrounded by a garden that is far from tidy. Egalitarians, unlike hierarchists and individualists, are not brand loyal, nor are they fashion conscious, nor do they wash their clothes on the same day each week, or sort them out into separate piles before they put them into the washing machine. Already, we can begin to see the power and environmental relevance of this sort of approach. For instance, if you are thinking of siting a nuclear waste repository somewhere that is geologically ideal, and the people round about turn out not to sort their washing out into separate piles, forget it!
- *Fatalists* make few long-term plans, find it difficult to save money and are much addicted to take-away meals. They do not have regular routines or allotted household tasks or their own places at the table. Indeed, their take-away meals are likely to be eaten in front of that quintessentially fatalist piece of technology – the television set (Putnam, 1995; Schmutzer, 1994).

So it looks as though culture, in the sense of these different solidarities that, in their varying strengths and patterns of interaction, make one national (or organizational) culture different from another, does matter. And the implications of this realization are rather serious.

Some Implications

Most current research on consumption is focused entirely at the macro-level, with the micro-level being handled in terms of *per capita* (or per household) *consumption* which, of course, is just one macro-number (national consumption) divided by another macro-number (national population, or national population divided by average household size). This is consistent with non-cultural approaches (which, if they are not based on the non-satiety requirement, tend to assume that needs are standardized and are set by a country's level of economic development) and also with those cultural approaches which assume that what matters is *national* culture. Cultural Theory, however, points out, first, that these latter approaches are treating culture as an uncaused cause ("Why do the Chinese do that?"; "Because they're Chinese";) and, second, that they are ignoring the cultural plurality *within* each nation: the social solidarities which, by their differing proportions and patterns of interaction, actually make national cultures different from one another.

The much relied-upon notion of national *per capita consumption* becomes statistically invalid if each nation is comprised of four (more properly, five) distinct consumption populations. Indeed, it makes more sense to aggregate each consumption population *across* nations, rather than lump them all together as American consumers, Chinese consumers, Indian consumers and so on. This argument, if valid (and the study of British household consumption styles strongly suggests that it is) has some profound consequences: the writing-off, for instance, of pretty well all the present modelling work, together with all the policy decisions that have been justified by that work.[14]

With a moratorium on the use of national per capita consumption (who, after all, would want to be caught committing the statistical sin of homogenizing heterogeneity?), attention can then shift to the styles: their impacts on one another and the possibility of changing some of the items that at any time are defining those styles (they're changing all the time anyway) for others that are less environmentally harmful – although what is considered to be environmentally harmful will vary with the social construction of nature that accompanies and supports each solidarity. For instance, those who are not egalitarians are unlikely to be convinced that they need "a whole new relationship with nature",[15] but they can readily latch onto the desirability of eating lower on the food chain. Many of the highest earning and best educated Americans are doing just that, not (if they are individualists) to save the world but in the pursuit of healthy living and personal success. At the same time, of course, many of the poorest and least well-educated Americans have moved themselves (or, in the case of that high proportion who are fatalists, have found themselves moved) up the food chain. All of which suggests that education and productive employment – fatalism reduction, in other words - would be worthwhile policy goals in relation to global climate change (and all the environmental and poverty-alleviation concerns that are wrapped up in that enormous issue).

This is not a facetious suggestion. The multiplier effect – it takes 30 kilograms of grain to produce just one kilogram of meat – means that a quite small shift down the food chain, in America (say), translates into a massive (and carbon sequestering) global shift in land use and land cover. Much the same holds for other changes in behaviour that the members of the various solidarities can pick up from one another: choosing a house, next time a person moves, that is nearer the workplace than the present house, for instance, or letting the forest grow back on a Scottish hillside instead of keeping it as a bald monoculture optimised for deer stalking. BMW funding a study of how to reduce the number of private cars in the centre of Munich is another example of this sort of constructive interplay: a *culturally plural and responsive citizenry interacting with reflexive policymaking so as to find better ways*

---

14 The exceptions (which include the perspectives approach in integrated assessment modelling pioneered by van Asselt and Rotmans, the "battle of the perspectives" explored by Janssen, and the sort of re-embedding of computer models in varying social contexts that is central to the ULYSSES Project) are listed in Rayner, Malone and Thompson (1999, pp. 38-39).

15 Ecocentrism instead of anthropocentrism.

of living together in a particular locality – security enhancement, in other words.[16] This is not to say that there is no role for global-level actors. Of course there is, but the anti-homogenizing lesson is best learned at the local level and then transferred to the global. It is, moreover, a lesson that, in line with the cultural method's rejection of the "veto on comparison", can be learnt anywhere in the world. In India, for instance, we find much the same heterogeneity of consumption styles as is evident in Britain (and I should add that I am indebted to Dipak Gyawali for the following analysis).

Consumption Styles in India

Gadgil and Guha (1995) have focused on the consumption of natural resources in India and have identified three distinct populations: the *omnivores* (the development-aided class of modern consumers), the *ecosystem people* (the traditional subsistence farmers and fisherfolk) and the *eco-refugees* (those who have neither the social contacts nor the entry fee to join the omnivores and, at the same time, have been unable to maintain their viability as ecosystem people). The omnivores, clearly, belong to the individualist solidarity, the ecosystem people (somewhat idealized) belong to the egalitarian solidarity, and the eco-refugees are the excluded fatalists. The interactions of these three solidarities, thanks to the interventions of the hierarchy (national government and international development assistance) are not constructive. Indeed, what we have in India, according to Gadgil and Guha, is almost the opposite of what has been happening in virtuous Munich.
They identify six root-causes of this unconstructive interaction:

- The ecosystem people are being deprived of the natural capital on which they depend.
- The ecosystem people are denied access to human-made capital by virtue of their not being much involved in the formal economy.
- The process of building human-made capital is itself inefficient and destructive of natural capital.
- The omnivores, unlike Adam Smith's market actors, do well even when others do not benefit. This is because their consumption of natural capital is subsidized by the state.
- Because of the omnivores' monopoly over human-made capital, the ecosystem people and the eco-refugees have no incentive to invest in the quality of their offspring; only quantity brings them any benefit.
- The concentration of human-made capital in a few urban centres (where it is fuelled by imported technology) has led to the mining of natural capital elsewhere, a process that is exacerbated by the omnivores dumping their wastes on those who are not omnivores.

---

16 For a detailed account of the cultural plurality that has characterized decision making in Munich, in contrast to the "monocultural hegemony" that has prevailed in Birmingham, see Hendriks (1994) and (1999).

These two heterogenizing examples - one from the industrialized world (Britain), one from the less industrialized world (India) - share the same typology of social solidarities, and they begin the task of opening up the sorts of dynamics - sometimes virtuous, sometimes not - that are at work at the various scales: household, village, nation, and so on. Together, they suggest that policy mono-cultures – approaches that insist that people are all the same (all insatiable, all with the same basic needs, or whatever) – are not what policymakers need for dealing with environmental (and other) problems. Indeed, they are what policymakers don't need. The idea of consumption as a moral activity – a way of supporting and strengthening a social solidarity – is therefore the first essential in getting to grips, in a useful way, with human needs and wants.[17] And it has been my argument that it is only by following the three rules of the cultural method that we can take account of the fact that culture matters, and from there move in a valid way to the consideration of environment across cultures (which, of course, is what we will have to do if we are to do anything sensible about climate change).

4. References

AWH, R.Y., 1976: Microeconomics: *Theory and Applications*, Santa Barbara, California: John Wiley.

Chapman, G.P. and M. Thompson (eds.), 1995: *Water and The Quest for Sustainable Development in The Ganges Valley*. London: Mansell.

Dake, K. and M. Thompson, 1993: The meanings of sustainable development: household strategies for managing needs and resources. In Wright, S.D., T. Dietz, R. Borden, G. Young and G. Guagnano (eds.) *Human Ecology: Crossing Borders*. Fort Collins CO: The Society for Human Ecology. pp. 421-436.

Dake, K. and M. Thompson, 1999: Making ends meet, in the household and on the Planet. *GeoJournal* 47,3. pp. 417-424.

Douglas, M. and S. Ney, 1998: Missing Persons: *A Critique of Personhood in the Social Sciences*. Berkeley and Los Angeles: University of California Press.

Eckstein, H., 1997: Social science as cultural science, rational choice as metaphysics. In Ellis and Thompson (op cit.). pp. 21-44.

Ehrlich, P. and J. Holdren, 1974: Impact of population growth. *Science*, 171: pp. 1212-7.

Ellis, R.J. and M. Thompson (eds.), 1997: Culture Matters: *Essays in Honor of Aaron Wildavsky*. Boulder CO and Oxford: Westview.

Gadgil, M. and R. Guha, 1995: Ecology and Equity: *The Use and Abuse of Nature in Contemporary India*. London: Penguin.

Hendriks, F., 1994: Cars and culture in Munich and Birmingham: the case for cultural pluralism. In D.J. Coyle and R.J. Ellis (eds.) *Politics, Policy and Culture*. Boulder CO: Westview

Hendriks, F., 1999: *Public Policy and Political Institutions: The Role of Culture In Traffic Policy*. Cheltenham; Edward Elgar.

---

17 An argument that is developed forcefully (and at length) in Douglas and Ney (1998).

Ives, J.D. and B. Messerli, 1989: *The Himalayan Dilemma: Reconciling Development and Conservation*. Routledge: London.

Linnerooth-Bayer, Joanne, 1999: Climate Change and Multiple Views of Fairness. In Ferenc L. Tóth (ed.) *Fair Weather? Equity Concerns in Climate Change*. London: Earthscan. pp. 44-64.

Putnam, R.D., 1995: Tuning in, turning out: the strange disappearance of social capital in America. *Political Science and Politics*. December. pp. 664-683.

Rayner, S. and E.L. Malone (eds.), 1998: *Human Choice And Climate Change* (4 vols.). Columbus, Ohio. Battelle Press.

Rayner, S., E.L. Malone, and M. Thompson, 1999: Equity issues and integrated assessment. In Ferenc L.Tóth (ed.) *Fair Weather? Equity Concerns in Climate Change*. London: Earthscan. pp. 11-43.

Schmutzer, M.E.A., 1994: *Ingenium und Individuum. Eine sozialwissenschaftliche Theorie von Wissenschaft und Technik*. Vienna and New York: Springer Verlag.

Schwarz, M. and M. Thompson, 1990: *Divided We Stand: Redefining Politics, Technology and Social Choice*. Philadelphia University of Pennsylvania Press.

Thompson, M., 1982: The problem of the centre: an autonomous cosmology. In Mary Douglas (ed.) *Essays In the Sociology of Perception*. London: Routledge and Kegan Paul. 302-328.

Thompson, M., 1998a: Waste and Fairness. *Social Research* 65,1. pp. 55-73.

Thompson, M., 1998b: Style and scale: two sources of institutional inappropriateness. In M.Goldman (ed.) *Privatizing Nature: Political Struggles for the Global Commons*. London: Pluto. pp. 198-228.

Thompson, M., R.J. Ellis, and A. Wildavsky, 1990: *Cultural Theory*. Boulder CO and Oxford: Westview.

Thompson, M., R.J. Ellis, and A. Wildavsky, 1992: Political Cultures. In M. Hawkesworth and M. Kogan (eds.) *Encyclopaedia of Government and Politics*. London: Routledge. pp. 507-520.

Thompson, M. and S. Rayner, 1998: *Cultural discourses*. In S. Rayner and E.L. Malone (op cit.) pp. 265-344.

Thompson, M., M. Warburton, and T. Hatley, 1986: *Uncertainty On A Himalayan Scale*. London: Ethnographica.

Tranvik, T., M. Thompson, and P. Selle: *Doing Technology* (and Democracy) The Pack-Donkey's Way: *The Technomorphic Approach* To ICT Policy. Oslo: Norwegian Power and Democracy Project. Report No. 9. (ISBN 82-92028-09-9).

Vidal, J., 1995: Nepalese hail move to scrap huge dam. *The Guardian* (London), 5 August, p.12.

The Legacy of Twenty Years of Energy Demand Management: we know more about Individual Behaviour but next to Nothing about Demand

Harold Wilhite
Elizabeth Shove
Loren Lutzenhiser
Willett Kempton

1. Introduction

Demand-side management (DSM) replaced energy conservation in the mid-1980's as the umbrella term for the science and policy of reducing the demand for energy. The semantic evolution has continued in recent years to "market transformation" and "energy efficiency on the demand side." This paper will argue that the nature and causes of "energy demand" have been oversimplified, reduced or ignored in the community of energy research and policy. Energy-related social science has largely been limited to the "behaviour" of the "end user." As a result, we do not know much more about the nature of energy demand today than we did in 1980. While there have been significant gains in energy efficiency over the intervening 20 years, the fact remains that the total energy demand in the United States and most European countries has increased (and in most cases has increased per capita as well). At the same time, the necessity for absolute reductions of fossil fuel use in industrial countries is more important than ever, given climate change and CO_2 reduction agreements.

In section 2, we look back at the ways in which energy social science was founded and the roles it was initially given in energy conservation research and policy. We trace some of the early difficulties with the dominant techno-economic approaches to reducing consumption and the discovery that the social sciences could be useful in addressing questions about (often unpredictable) human "behaviour" that might improve energy demand models. In section 3 we discuss the implications of this "behaviour" orientation and its consequent diversion of attention away from other important characteristics of societal demand. In section 4 we outline a new social science agenda which fills some of the gaps in the theories of energy demand which inform policy. We recast the demand for energy, and the things which use energy, as a social demand, dependent not just on prices and degree of consumer awareness, but also on social norms and a network of social institutions. We challenge energy research and policy to seriously address increasing demand for energy services.

E. Jochem et al. (eds.), Society, Behaviour, and Climate Change Mitigation, 109–126.

2. Energy and the social sciences

Following the 1970s energy crises, social scientists in the U.S. and elsewhere began to take an interest in energy as a social problem – paying particular attention to the impacts of energy shortages for various social groups (e.g., Newman and Day, 1975, Morrison, 1978, Dillman *et al.*, 1983). These studies drew on a long-standing interest in the environmental and technological foundations of human society by sociologists, anthropologists and human ecologists (e.g., Cottrell, 1955; White, 1975; see Rosa *et al.*, 1988 for a review). But while classic work in this area was largely theoretical, the new empirical studies were motivated by immediate policy problems and mainly concerned persons' attitudes toward the conservation of energy in short supply, patterns of energy use, willingness to change energy-using practices, and hardships experienced as a result of rising energy costs. Researchers came from the ranks of social psychology, anthropology, sociology, political science, and related disciplines. Their work ranged across multiple levels of analysis, including the individual, small group, community, firm, and society (see Lutzenhiser, 1993 for a review). Their results were published in a variety of social science and energy policy journals, and by the early 1980s there had been a sufficient growth in knowledge to warrant the formation of a U.S. National Research Council panel on energy and human society. The panel produced a widely-read overview volume entitled *Energy Use: The Human Dimension* (Stern and Aronson, 1984) and called for an expansion of social research on energy. Subsequently, there has been a steady decline in social science interest in energy per se (Lutzenhiser, 1992).

2.1. DEVICE-CENTERED APPROACHES TO UNDERSTANDING ENERGY USE

During that same period of the early 1980s, analysts working in universities, utilities, national laboratories, and government energy programs were taking a different approach to energy use. Trained in engineering and physics, and with some awareness of economics, they focused on machines, devices (e.g., furnaces, motors, lights, water heaters, air conditioning compressors, etc.) and buildings as "users" of energy. Elaborate mathematical models designed to mimic the "performance" of buildings and equipment – both individual devices and structures, as well as large populations or "stocks" of buildings and appliances – were constructed and used to estimate the effects of energy conservation initiatives, to assess the impacts of device-by-device efficiency improvements, to predict future changes in aggregate energy demand, and to explore the effects of policy on alternative societal level energy usage patterns (see Lutzenhiser, 1993 for a review of this approach and critique of its underlying paradigm). While there has been a significant decline in social science research on energy use and conservation since the 1980s, work informed by such device-centered approaches has continued and grown.

Trained in technical disciplines, technical modelers quite naturally see the energy problem as one involving flows of energy (electrons, natural gas, petroleum) through physical systems that convert it into heat, light, motive power, etc. The energy use rates of these systems, in principle, can be closely estimated and improvements in efficiency – an obvious conservation strategy – can be assessed. In initial thinking about these systems, humans were believed to be of little importance, in the sense that they were simply recipients of energy services and thereby not an important part of the energy-using system subject to analysis. However, efforts to match models with measurements of real world energy flows turned out to be problematic. Humans re-appeared as active energy "users" – manipulating devices, managing buildings and interacting with energy flows at every turn. This realization led some physical modelers to invite social scientists to join in an effort to improve the predictions of device-centered modelling systems.

2.2. BRINGING THE SOCIAL SCIENCES TO BEAR ON ENERGY POLICY RESEARCH

The carving out of a "behavioural" niche in energy research for the social sciences in the early 1980's came in response, then, to the sometimes dramatic problems encountered in applying technical and economic perspectives to energy policy. One early study, for example, showed that physically-identical townhouses varied by 2-to-1 in energy use, presumably due to occupants (Socolow, 1978). A related analysis by Sonderegger showed that about half the variation among similar houses was due to the occupants, concluding "We have proved experimentally that (so far) unpredictable behaviour patterns of the occupants introduce a large source of uncertainty in the computation of residential space heating energy requirements ... there is little practical usefulness in pushing too far the detail of any deterministic model [e.g. physical/engineering model] for the prediction of heating load requirements... (1978: 323)." For apartments in moderate climates, where appliance use was a larger proportion of the total, variation in total energy use among identical apartments was found to be as great as 10-to-1 (Diamond, 1984). Amendment of technical models with social variables was found to significantly improve their ability to estimate measured consumption in the real world (Cramer *et al.*, 1985).

These findings opened a space for the social sciences in the energy policy and program arenas. Social psychology was the first discipline to occupy that niche, bringing with it perspectives on individual motivation and information (Stern and Aronson, 1984; Ester 1985; and deYoung 1996). Another early concern was with the "diffusion" of technologies. Here social psychologists and sociologists attempted to understand why persons were not adopting new, more efficient, devices at predicted rates. This "failure to adopt" flew in the face of assumed consumer (and business) rationality, since the economic rewards of more efficient technology clearly outweighed their costs. Some anthropologists and sociologists pointed to the problems of applying diffusion theory when dealing with household rather than individual decision making (Kempton and Montgomery, 1982; Wilk and Wilhite, 1985). This turn of attention to households proved fruitful, subsequently opening analysis to the consideration of socio-cultural differences among groups in levels of energy consumption as well as in understandings, willingness to conserve, and ability to make technological changes (Hackett and Lutzenhiser, 1991; Cebon, 1992; Lutzenhiser and Hackett, 1993; Erickson 1997).

Efforts to bring about a "social amendment" of the dominant paradigm of policy research have not been very successful. There has been little interest in the U.S. in improving upon the device-centered model, and while European attempts have been more ambitious (e.g., IAE, 1995; ECU, 1996) they have also had a limited impact upon conventional theory and practice.

2.3. A LIMITED SOCIAL SCIENCE CONTRIBUTION

Although leading energy analysts understand that human action is the central and controlling element of energy systems (e.g., Schipper, 1991; Lovins, 1992), the insight is not widely acknowledged in energy policy discourse. As noted, social science interest in energy has declined since the mid-1980s, and several important developments have contributed to researcher disenchantment with the energy policy agenda. These include the rise of DSM, more pressing concerns about global environmental change (GEC), and recent energy system interest in "market transformation" (MT).

DSM translated energy "conservation" into "efficiency," and "efficiency" into "least cost source of energy supply," forcing a narrow policy focus on the marginal costs of small improvements in devices. In the DSM approach, humans are of interest only as "free riders" in utility subsidy or supply acquisition schemes. The second development, GEC, captured the attention of social scientists who had been concerned about energy in the 1980s – and who had even been involved in efforts to improve the device-centered approach – but had grown weary of the limited role represented by the terminology of "human factors". By contrast, in the evolution of GEC thinking in the 1990s, energy came to be seen as one of several crucial variables involved in the "society-environment interactions" responsible for environmental change on a global scale (Stern *et al.*, 1992). From the vantage point of GEC, DSM seemed (and still seems) a narrow and self-absorbed enterprise. The third development, MT, could result in a renewed interest in the contributions of the social sciences, since it views the task of efficiency as one which involves less costly intervention in markets to induce the diffusion and adoption of more efficient devices, buildings, etc., but again, the human role is reduced to one of "behaviour" in physical systems – in this case behaviour that might be shaped through various "marketing" efforts. In MT there is also a continued focus on the "individual" (and continued confusion about the role of larger forms of organization), as well as an unfortunate and misleading language of "market barriers," "market interventions," "exit strategies," and so forth, which puts markets at the center and social context at the periphery.

The inability of social scientists to bring insights to bear on energy and environmental problems can be partly traced to the political and organizational contexts of energy research agenda-setting. For a variety of reasons (see Lutzenhiser and Shove, 1999), those agendas are stubbornly reliant – despite two decades' evidence of poor performance – on a view in which human "behaviour" remains conceptually distinct from the workings of devices, buildings, infrastructures and the other socio-technical arrangements involved in energy use. In the balance of this paper, we consider the continuing limitations of this conception, argue for a more social view of energy use, and offer an expanded research agenda that addresses the interests of the GEC and DSM/MT communities in reducing the amount of energy used to sustain social life.

3. The problems of reducing the role of social science to that of understanding the end-user's "efficiency" behaviour

Most of the early programs and conferences which invited social scientists in as participants in energy conservation research and policy analysis were centered around understanding, or accounting for "behaviour". One of the early fora responsible for opening energy research to social science was the American Council for an Energy Efficient Economy (ACEEE) and its Summer Studies on Energy Efficiency in Buildings (an intensive one week conference for researchers, policy makers and energy practitioners with a number of parallel sessions on topics surrounding energy-efficiency technical issues, policy, programs and program evaluation). A watershed was passed in 1980, when the organizers recruited social psychologists from the university local to the conference location (UCSC) to organize and add a "behaviour" panel to an otherwise heavily technical conference. The every-other-year ACEEE Summer Study provided the basis for a continuing, if minority, interest in energy-related social science research. Tracing the changes in the title of this panel sheds light on the changing role of social science over the past two decades.

By the late 1980's the panel title "Behaviour" was changed to "Behaviour and Lifestyle", partly in response to the work by Morrison and Schipper on lifestyles (Morrison, 1979; Schipper *et al.*, 1989). This concept of "lifestyle" lives on in many energy circles, though divergent operative definitions abound. While Schipper and his co-authors called for the addition of demographic elements to the definition of lifestyle, most follow the conventions of market research, defining lifestyle clusters by aggregating individual beliefs, values or attitudes (Shove *et al.* 1998). Lifestyle was dropped from panel title in 1990 partly in response to concern that the terminology failed to capture the full range of actors, practices and interactions relevant to the understanding of energy consumption.

The 1990 title, "Human Dimensions," maintained over 4 Summer Studies (8 years), was thought to be broad enough to subsume behaviour, lifestyle, marketing and studies of diverse "end users", including institutions. However, in 1998 there was a reversion to the title of Behaviour, thrown together in an odd mixture with information technologies and non-energy benefits.

This story of the labeling of the social science role indicates tentative steps towards a more comprehensive social approach to understanding energy consumption. However, the reinstatement of the behavioural terminology suggests that the underlying approach has not changed much over the last 20 years. In the remainder of this section we explore the problems and consequences of this narrow focus.

3.1. THE INDIVIDUAL AS THE LOCUS OF CONTROL AND CHANGE

In an article in Per Otnes' (1988) anthology *The sociology of consumption*, he takes us through a typical day in his life, getting out of bed, brushing his teeth, showering, shaving, making coffee and toast, driving to work, and so on. He then goes on to discuss how each of these activities is bounded by a limited matrix of choice from the point of view of the end consumer (deemed "structures of constraint" by Kabeer (1994)). The point is that individual choice in industrial societies is limited by the way cities, energy and water supply systems, housing designs, product designs, etc. are configured. Individuals can influence what happens at the end of the pipe, but significant changes in energy use are bounded by the "upstream" systems they are plugged into. The way these systems come into place, the interactions among the actors involved in their construction and maintenance, and thoughts on how they might be changed are certainly relevant for the science of energy reduction. Yet the concept "behaviour" neither signals their importance nor captures their workings.

In short, if one accepts that significant changes in the ways we use energy will be predicated on a significant social transformation, then focusing on behaviour of individual end-users as the only key to change is both overly simplistic and counter-productive. As anthropologist Sidney Mintz (1979) put it over 20 years ago, there is a need "to specify with more confidence the way individuality plays itself out against terms set by socio-cultural forces." This is not to say that understanding individual action and choice is irrelevant. The point is that this has been taken as the only role for social science.

3.2. OVERCOMING THE BARRIERS

Above, we discussed the socio-psychological models that dominated early energy social science. Other models which have had a strong influence assume that people respond rationally to economic and technical opportunities (call it techno-economic rationality). One of the main roles of social science has been to figure out why people do not act in accordance with this model. For example, perhaps a particular energy-efficient device, if purchased, would yield energy savings, and thus an provide an "income stream" better than alternative investments such as bonds or stocks. Given this definition of the problem from economics or technology, the social scientist is offered the task of figuring out why energy users are not adopting the technology (even working within this paradigm, energy users can be seen as rational, but optimizing other things than what physical or economic models predict e.g. see Kempton 1986 on folk theories of thermostats). Thus social science has been diverted into the exercise of barrier analysis. This is at best a limiting role which distracts resources and attention from the effort of understanding changes at the societal level, for instance, in the conventions and norms of comfort, aesthetics or convenience.

3.3. ENERGY CONSUMING BEHAVIOUR OR THE CONSUMPTION OF ENERGY SERVICES?

Aside from those who deal with the production or transmission of energy, no one really "behaves" in relation to energy (with the exception of someone who inadvertently sticks their finger in the electricity outlet). Devices convert energy into services; people are interested in services, not energy. So what is needed is a social science of energy service consumption, something much broader than a science of energy behaviour.

The recent literature abounds with calls for researchers and policy makers to move the focus from energy to energy services (see Wilhite *et al.*, 1996a; Wilhite and Lutzenhiser, 1998; also Laitner *et al.* in this volume). The services most often discussed are space heat and cooling, light, and cleanliness (or hygiene). Understanding how and why demand for these services is growing is essential in order to making any headway on developing instruments for encouraging reductions in energy use. We would argue however that the broadening of the frame from energy to service has thus far been modest. There are developments driving energy demand which have not yet been classified as energy services at all, examples being escalating dwelling size, comfort and convenience.

Dwellings everywhere in the industrial world are getting larger. The number of specialized rooms is growing (offices, second baths, individual bedrooms for children, entertainment rooms, etc.), as are the number of appliances, and the physical space needed to accommodate them, particularly for entertainment, information and kitchen services. Bigger volume means larger space heat and cooling needs, two of the strongest drivers of energy demand. An example is the growth in energy used to heat Norwegian homes. In spite of more stringent building codes and a doubling of the thermal efficiency of the Norwegian home from 1960 to 1990, energy use for space heating rose over the period, due to a doubling of the per capita size of dwellings (Hille, 1997). Central space heating is not common in Norway, but indications are that in other OECD countries, the transition to central space heating has contributed significantly to increased energy use. The share of central space heating in both Denmark and France doubled in the period from 1960 to 1980, tripled in Germany and Italy, and increased eight-fold in the Netherlands. The use of energy for space heat has more than doubled in all of those countries over the same period (Chateau and Lapillone, 1980). Another example is space cooling in Japan, where air conditioners are technically very efficient, but where space cooling demand is still increasing dramatically, due to increases in dwelling size, changing tastes and modern building designs which do not support natural cooling (Wilhite *et al.* 1996b). All of these dimensions - size, taste and design - are crucial to understanding energy consumption.

Comfort and convenience can be thought of as meta-energy services. In the literature on energy behaviour and lifestyles, their importance has been acknowledged in analyzing how and why, and for what purposes people use energy. The issue which has yet to be addressed relates to the dynamics of change: How do notions of comfort and convenience evolve, and how does this relate to the provision and consumption of energy?

Changes have been dramatic in almost every aspect of energy consumption: indoor space cooling and heating; clothes and dish washing; bathing; media entertainment; and more. Cooper's (1998) story of how America was air-conditioned documents the transformation of comfort standards. In the early days of air-conditioning in the 1920s, temperatures around 25 °C were considered comfortable. The first ASH and VE guidelines published in 1925 were for 25.5 °C and 50% relative humidity. Building operators in the United States now view 21 – 22 °C as being an appropriate indoor summer temperature, and some service establishments such as malls and theaters set temperatures as low as 19 °C. The increasing use of hot water as a provider of pleasure (hot tubs, jacuzzis, Roman-sized bath tubs), and space heat and light to provide atmosphere, are other examples of rapidly evolving ways in which energy provides comfort.

Regarding the notion of convenience, Warde *et al.* (1999) argue that convenience has become an obsession in modern industrial societies. In fact the authors define the "hypermodern" society to mean the convenience-obsessed society. They show how convenience in the 20th century meant "saving" time, i. e. compressing the amount of time taken to do an activity like clothes washing. More recently the concept has expanded to include the "scheduling" of time, the process of ordering and managing activities so as to be able to cram ever more events into a given day or week. Both of these aspects of convenience have implications for energy use, the former justifying the introduction of ever more efficient mechanical substitutions for manual work, the latter leading to more activities, more devices to manage the pressures of time, more traveling and greater demand for faster means of getting from one place to another.

Erickson (1997) looked at the "scheduling" aspect of convenience in her cross-cultural ethnographic study of household energy use in Sweden and Minnesota. In both places the respondents indicated that being "busy" was an important indicator of a successful life. At the same time "busyness" was said to be one of the greatest contributors to stress and dissatisfaction (this apparent paradox of striving for something and at the same time being distressed by it is also addressed by Juliet Schor in *The Overspent American* (1999)). In a newly published anthropological study of "Silicon Valley" in California, reported in Knowltan (1999), the researchers found that people used an inordinate amount of time in "making their busy lives manageable." People equip their homes with the latest gadgets so they can work at night and get ahead (or as one interviewee put it "to stop me from falling behind"). They use voice mail, modems and digital organizers to exchange schedules among family members and to divide up domestic chores. As one of the researchers, Charles Darrah, observed, "It's not so much a life lived, but a life managed." Silicon Valley is not representative of your average community, but the idea of the "managed life" may be spreading. It signals a change in the notion of convenience that could lead to significant increases in energy consumption.

Studies documenting changing societal demands for comfort and convenience touch on the classic debate in the social sciences regarding "needs" and "wants." Energy analysts have drawn distinctions between "basic" and "other" energy needs. Viewed in the context of evolving energy services, these distinctions are moving targets which are analytically problematic. Once established and reified, the categories divert attention and debate from underlying inequalities in access to energy. This is true for privileged as well as underprivileged groups. Is an automobile a "need" or luxury for someone without access to a well developed and well functioning public transport system? Is central air conditioning a luxury or necessity in a building without good natural ventilation? These are hopeless questions. They disguise the political issues surrounding distribution and equality and impede the discussion of important theoretical issues surrounding the growing social demand for energy and how wants are constructed and manufactured. We agree with Slater (1998) that the issue is not about appropriate "levels" of energy use, but about how such levels are conceptualized (deduced, analyzed and debated). We must go beyond merely descriptive accounts of changing expectations of convenience and comfort, important though these are. What the energy world requires, and what social scientists are in a position to provide, is an analysis of how conventions evolve, of how energy-intensive ways of life become normal, and of how energy demand is embedded in society.

4. Demanding a new agenda?

In this section we outline a framework for social science research which addresses the challenge of understanding escalating demand for energy-intensive services, practices and ways of life and which goes beyond a narrowly behavioural perspective.

At first sight, the challenge of understanding behaviour might not seem to be so different from that of understanding demand. In the world of energy policy, the behaviour which is of interest, and which has been the focus of attention, is that which is associated with energy consumption and hence with demand for energy resources. In terms of social theory these two concepts are, however, a world apart. As we have shown, theories of behaviour have an ancestry grounded in psychology and the study of individual belief and action. By contrast, the concept of demand points to the development of markets, the social and technical construction of needs, and the steady evolution of expectations about what constitutes a 'normal' way of life.

Economics appears to offer a bridge between these two strands: notions of rational action, market imperfections and the barriers to technical change and innovation all promise to link macro trends at the level of markets to the micro level patterning of individual behaviour. Though it is an important element in this narrative, theorizing about demand remains relatively undeveloped: in energy planning, at least, the working assumption is that there is demand for energy and that the nature of demand is sometimes influenced by price (Boardman 1991). Such an approach generally involves distinguishing between needs and wants; that is, between "necessary" and "optional" energy demand, then speculating on the implications of this distinction for consumer responses to different price signals. But that is about as far as it goes.

4.1. RE-POSITIONING DEBATE

In shaping up a new agenda we begin by re-framing the discussion of needs and wants. Rather than spending further effort trying to distinguish between the two we take a different tack, asking how conventions of social life come to be established and what this means for energy demand. In so doing we make three critical steps.

First we take the escalation of energy demand and the evolution of consumer expectations as a problem to be explained and understood in social, cultural and collective, rather than individualistic, terms.

Second, we view costs and prices as a secondary dimension. Recognizing that economics is a relative not an absolute enterprise and acknowledging that values are anchored in social judgement, our goal is to understand how energy consuming practices come to be valued as they are. In this we agree with Hefner's (1983) critique of neo-classical economic approaches to consumer choice: "most neo-classical studies do not investigate the history or social genesis of subjective preference...An adequate account of economic action...requires examination of the interaction between individual preferences, the means available for their satisfaction, and a social world which shapes both." By asking how demand is made, constructed and sustained, we are drawn, like it or not, into an analysis of the inter-dependent practices of producers, providers, utilities, and governments.

Our third and perhaps most important change of emphasis is to acknowledge that energy is not a meaningful term when it comes to understanding consumption and demand. People do not consume energy. They consume the services it makes possible. This is a simple but important point. Graphs of increasing energy consumption are, in fact, graphs of the societal appropriation or increasingly intensive use of technologies such as lighting systems, refrigerator-freezers, air conditioning, and so on. If we are to understand the dynamics of energy use, we are consequently obliged to take note of the development and diffusion of energy using devices and technologies. Understanding the dynamics of energy demand (or demand for the services energy makes possible) is thus an exercise in understanding socio-technical change and the co-evolution of infrastructures, devices, routines and habits.

This approach has immediate implications for the framing of problems and for the sorts of questions which are, and are not, important to address. The table below highlights some of the assumptions and starting points which underpin the traditional, behavioural agenda and the more challenging analysis of demand which we advocate.

Having established the principles and starting points for a new agenda, we are now in a position to identify some of the research questions which emerge, and consider the range of theoretical and methodological resources which might be used in response.

Table 1: Established and new agendas for a social science of energy demand

Understanding energy behaviour	**Understanding the construction of demand**
Choices are driven by economics	Economics is relative (changes through history) and contextual (embedded in other systems of decision and desire)
Separation of energy consumption from the analysis of other forms of consumption	Social scientific theories of consumption are relevant for understanding energy
Demand comes from consumers; hence the focus on consumer choice	Producers and consumers are implicated in the co-evolution of demand, and the choices of both are highly structured
Consumer choice is sovereign	All of the forces contributing to escalating demand for energy services may be studied for policy opportunities
Focus on classic end uses: lighting, heating, cooking, etc.	Focus on changing conventions of comfort, cleanliness and convenience.
Demand side management viewed as a bounded technique for influencing behaviour	Recognition that demand is a societal, not an individual phenomenon; it could be managed at multiple levels
Invokes a distinction between needs and wants, and assumes that the latter are subject to individual preference	Rejects the relevance of needs-wants distinctions on the grounds that for all practical purposes both are socially constructed

4.2. RE-FORMULATING QUESTIONS

Having turned the debate around it is clear that the primary challenge is to understand the dynamics of demand. How are new 'needs' constructed, and how do expectations of comfort and convenience evolve?

One means of addressing this question to document change and variation either historically or cross culturally. This does not on its own advance theoretical understanding of the processes involved and in any event, we might expect different dynamics to be at play in different contexts and circumstances. Nonetheless there is important work to be done by taking the practice of everyday life as the starting point and seeking to capture the variety of ways in which 'normality' is made and reproduced. Comparative research undertaken as part of a European Science Foundation-funded initiative on "Consumption, Everyday Life and Sustainability" revealed the value of such an approach, examining 'normal' energy consuming practices in Turkey and Denmark, and tracking the recent evolution of both (Ger *et al.* 1997).

Of course there is enormous variation in what constitutes normal standards of comfort, cleanliness and convenience (and so energy demand) within any one society. More needs to be known about how expectations change between one generation and the next, how they differ between social classes and what this means for the socially mobile.

We have already learned that the management of demand for energy (in terms of its extent and timing) involves many actors: builders, utilities, estate agents, government regulators, retailers, and engineers are all implicated in the construction and maintenance of normal energy consuming practices. Some of the more successful energy efficiency-oriented programs have taken a more integrated approach to connections and interrelationships among these various actors. Examples are Energy Star[1] in the USA and the market transformation programs of the Swedish National Energy Administration (STEM)[2]. Understanding the nature of the various interests, and the relationships between them with respect to different fields of consumption, such as the indoor environment, laundry or food provisioning etc. represents a further challenge. Better knowledge of differences in the organization and management of energy demand at home and at work also promises to illuminate the different ways in which similar services might be provided.

Recognizing that we inhabit a total regime of energy consumption prompts questions about the distinctive dynamics of the manufacturing of demand at different levels, and about how the possibilities and opportunities open to households relate to the structuring of national systems and infrastructures. It is also important to focus attention on moments and thresholds of change. Major investment now has implications for the structuring of future supply and demand, just as the practices of the past influence perceptions of energy use today.

We do not argue that research and policy which focuses on individual belief, action and choice is in itself irrelevant, nor do we suggest that more systemic research

---

1 Energy Star is a national program which assigns an energy star label to equipment which meets a given energy efficiency standard. A focus of the program is to encourage partnerships among relevant actors. According to Skip Laitner (personal communication) of the USEPA, the program now has more than 1200 manufacturers producing a total of 7000 individual product models in 30 consumer product categories.

2 STEM has heavily promoted technology procurement, which it defines as "a kind of public competition in which different types of actors are brought together to define a technology goal which is reached through a competition including market incentives" (Lund, 1997).

strategies should take the place of "human dimensions" style enquiry. But neither do we think these strands can be woven together to form some new integrated approach to energy research. An integrated understanding of societal energy consumption is one which views individuals and institutions in complex relations; in fact, the nature of these interrelationships provides the most fertile ground for new research perspectives on how energy services are evolving.

The paragraphs above outline themes and questions which deserve further attention. They illustrate the sorts of problems which would figure on a research agenda which really was about the management of energy demand. It is already obvious that the familiar repertoire of economic and social psychological expertise is not sufficient to cope with challenges of this kind, and that we need to make use of a much broader armory of theoretical concepts and methodological approaches. The following section identifies some of the resources available.

4.3. RE-TOOLING RESEARCH

What ideas can we borrow, appropriate and adapt in order to address the dynamics of energy demand? This necessarily speculative section identifies a handful of concepts which promise to be of real value, and identifies some of their limitations as well.

We have noted above that energy consumption is always mediated by technology. Following this theme, there is much to be gained from the sociology of science and technology. While more recent studies address regime-level transformation (De Laat, 1996; Schaeffer 1998), most work concentrates on innovation and change at the level of individual devices such as the bicycle (Bijker, 1995), the fluorescent light (Bijker, 1992) or the refrigerator (Cowan, 1985). The career of things has been the focus of attention rather than the services which they make possible, or the patterns of inter-dependence which link devices together to form co-dependent suites of technology (and of attendant practices). The units of analysis so far fail to include the kitchen or the home as a total energy system. Links are sometimes made between individual technologies and the infrastructures of which they are a part (e.g. Forty, 1986) but more could be done by way of understanding these relationships and their implications for regime level transformation.

Changing tack, the sociology of consumption has much to offer energy researchers by way of ideas about meaning, identity, acquisition and so on. For obvious reasons, there is relatively little work on what we might term 'inconspicuous' consumption (Shove and Warde, 1998) – for instance the consumption of water or electricity. On the other hand, Fine and Leopold's (1993) 'systems of provision' approach offers a very relevant framework with which to understand the structuring and organization of utility sectors.

Environmental sociologists have paid more attention to the use of energy than most, but again their interests are partial. The idea that environmentally conscious consumers might drive a move towards 'ecological modernization' and so transform the relationship between households and energy suppliers is contested and controversial. Nonetheless, the work of Spaargaren (1997) and others at least reminds us of the extent to which consumers are implicated in the design and operation of utility services.

Anthropological research reveals cross-cultural variation, demonstrates the malleability of normal conventions, and highlights the symbolic significance of energy consuming practices and habits. Social historians have made valuable contributions to documenting diversity and tracking change. In both cases, there is real potential for the development of more theoretical accounts of stability, order, routinization and innovation.

This brief and selective catalogue suggests that there are existing tools and resources which can be used to tackle the broader agenda of energy demand. It also suggests the need to develop, extend and adapt many of these ideas and approaches. There are gaps within and between sub-disciplines which threaten to hamper research in this field. This uneven intellectual backdrop reflects the fact that few sociologists, social historians, anthropologists or scholars of science and technology have devoted serious attention to the dynamics of energy demand. This is hardly surprising since the field has been described in terms of 'behaviour' and marginalized as a result.

If questions of demand are to be tackled head on, and if there is to be any chance of realizing the potential for exploiting theoretical and conceptual resources from the social sciences, it will be necessary to draw a new population of social researchers into the field. The present framing of research agendas in terms of 'end user' behaviour is not inviting.

4.4. RE-THINKING POLICY

Several policy implications follow from our argument. First, if device-centered approaches are limited in their ability to represent the real world conditions of consumption, then the policy initiatives guided by those approaches (e.g., with a focus on engineering design, provision of technical information, incentive subsidies for equipment purchase, etc.) will misunderstand social action and fail to deliver predicted energy savings, something that Wirl (2000) argues has been the case with DSM. Second, even with a "behavioural" amendment (e.g., technical assistance/ public information initiatives aimed at "decision-makers"), device-oriented models direct policy attention primarily to the behaviours of engineers and accountants, in the case of firms, or furnace-repairs and energy bills, in the case of households – in both cases, low status areas that are likely to receive little attention in decision-making, particularly when energy costs are small. Third, if social perspectives on demand offer a more accurate conception of energy use, then policy analysis is able to explore a larger number of causal accounts and entertain a wider range of interventions. This is particularly important for the development of policies that might successfully stimulate energy savings and significant CO_2 reductions. It is also important in uncovering inequalities in the present system and anticipating equity problems in proposed policies.

Given that energy use is shaped in complex systems that often submerge energy and other environmental concerns, it follows that existing policies not directed toward energy and the environment are nonetheless likely to shape patterns of energy consumption. Here we have in mind state policies in the form of subsidies for fuels, zoning and land use planning, transportation regulations, industry protection arrangements, building codes, engineering standards, etc. The effects of these sorts of existing policies ought to be thoroughly considered in policy analysis and the development of new energy saving initiatives.

5. Conclusions

Our intention in the paper has been to explore a new approach to the science of energy demand: one which adequately accounts for the actors, institutions and networks which contribute to change; which re-envisions the object of inquiry as the services which energy provides; and which is equipped to understand change. This new approach would not obviate the individual, nor research intended to track changes in how individual consumers think and act; it would, however, recast demand as the result of interactions in the social, cultural and technical contexts in which individual lives are played out.

6. References

Boardman, Brenda, 1991: Fuel poverty: from cold homes to affordable warmth. London : Belhaven Press.

Bijker, Wiebe, 1992: The social construction of fluorescent lighting, or how an artifact was invented in its diffusion stage. pp. 75-102 in W. Bijker and J. Law (eds.) Shaping Technology/Building Society: Studies in Sociotechnical Change. Cambridge, Mass.: MIT Press.

Bijker, Wiebe, 1995: *Of Bicycles, bakelite and bulbs: Toward a theory of sociotechnical change*: Cambridge Mass.: MIT Press.

Chateau, B. and B. Lapillonne (eds.), 1982: *Energy Demand: Facts and Trends*. Springer-Verlag: Vienna and New York.

Cowan, R. Schwartz, 1985: How the refrigerator got its hum in Mackenzie, D., and Wajcman, J. (eds.). *The Social Shaping of Technology*, Open University Press: Milton Keynes.

Cebon, P., 1992: Twixst cup and lip: organizational behaviour, technical prediction and conservation practice. *Energy Policy* 20: pp. 802-14.

Cooper, G., 1998: *Air Conditioning America: Engineers and the Controlled Environment*. The John Hopkins University Press: Baltimore and London.

Cottrell, F., 1955: *Energy and Society: The Relation Between Energy, Social Change and Economic Development*, New York: McGraw-Hill.

Cramer, James., Nancy Miller, Paul Craig, Bruce Hackett, and Thomas Dietz et al. 1985: Social and engineering determinants and their equity implications in residential electricity use. *Energy* 10, 12: pp. 1283-91.

de Yong, R., 1996: Some Psychological Aspects of Reduced Consumption Behaviour: The Role of Intrinsic Satisfaction and Motivation. *Environment and Behaviour* 28.

Diamond, Richard C., 1984: Energy use among the low-income elderly: A closer look. *Proceedings, 1984 ACEEE Summer Study on Energy Efficiency in Buildings*, pp. F-52 - F-66.

De Laat, B., 1996: *Scripts for the Future*, PhD thesis, University of Amsterdam.

Dillman, Don., Eugen Rosa and Joye Dillman, 1983: Lifestyle and home energy conservation in the U.S. *Journal of Economic Psychology* 3: pp. 299-315.

ECU, 1996: Workshop on Modelling Energy Behaviour, Environmental Change Unit, University of Oxford, Oxford, UK.

Erickson, R., 1997: *Paper or Plastic? Energy, Environment and Consumerism in Sweden and America.* Praeger: Westport.

Ester, Peter, 1985: *Consumer Behaviour and Energy Conservation.* Dordrecht: Martinus Nijhoff.

Ger, G., H. Wilhite, B. Halkier, J. Lessøe, M. Godskesen, and I. Ropke, 1999: Symbolic Meanings of High and Low Impact Consumption in Different Cultures, in Shove, E., ed., *Consumption, Everyday Life and Sustainability.* Centre for Science Studies, Lancaster University, England.

Fine, B. and E. Leopold, 1993: *The World of Consumption*, London: Routledge .

Forty, A., 1986: *Objects of Desire: Design and Society Since 1750*, London, Thames and Hudson.

Hackett, Bruce and Loren Lutzehiser, 1991: Social Structures and Economic Conduct: Interpreting Variations in *Household Energy Consumption Sociological F*orum 6: pp. 449-470.

Heffner, Robert W., 1983: The problem of preference: Economic and Ritual Change in Highlands Java. *Man* 18(4): pp. 669-689.

Hille, John, 1997: *Sustainable Norway.* Lokal Agenda 21, Idehefte. Idebanken: Oslo.

IEA., 1995: *Introduction of Cultural and Institutional Factors into Modelling Energy.* Geneva: International Academy of the Environment, Event Report E44.

Kabeer, Naila, 1994: *Reversed Realities : Gender Hierarchies in Development Thought.* W. Norton & Company: London.

Kempton, Willett, 1986: Two Theories of Home Heat Control. *Cognitive Science* 10: 75 - 90. Reprinted and expanded 1987 in *Cultural Models in Language and Thought*, D. Holland and N. Quinn (eds.), Cambridge University Press. Also appeared, abridged, in Morrison and Kempton (eds.), 1984 (see below).

Kempton, Willett and Laura Montgomery, 1982: Folk Quantification of Energy. *Energy--The International Journal* 7(10): pp. 817-828.

Knowlton, Brian, 1999: Silicon Valley: Living the Managed Life. International Herald Tribune, Paris, 6. October.

Lovins, Amory, 1992: *Energy Efficient Buildings: Barriers and Opportunities.* Boulder, CO: E-Source.

Lund, P., 1997: Evaluation of the Swedish programme for energy efficiency – successful examples of market transformation through technology procurement. *Proceedings of the 1999 ECEEE Summer Study.* European Council for an Energy Efficient Economy, Paris.

Lutzenhiser, Loren, 1992: A Cultural Model of Household Energy Consumption. *Energy-The International Journal* 17: pp. 47-60.

Lutzenhiser, Loren and Bruce Hackett, 1993: Social Stratification and Environmental Degradation: Understanding Household CO2 Production. *Social Problems* 40: pp. 50-73.

Lutzenhiser, Loren and Elizabeth Shove, 1999: Coordinated Contractors and Contracting Knowledge: The Organizational Limits to Interdisciplinary Energy Efficiency Research and Development in the U.S. and U.K. *Energy Policy* 27: pp. 217-227.

Mintz, Stanley, 1969: Comments: Participant-Observation and the Collection of Data, Vol. 4 of the *Boston Studies in the Philosophy of Science,* 1966-68: 341-49. Reidel Publishing Co.: Dordrecht, Holland.

Morrison, Bonnie M. and B. Long, 1979: Energy and Families: The Crisis and the Response. *Journal of Home Economics* 70 (5).

Morrison, Bonnie M. and Willett Kempton (eds.), 1984: *Families and Energy: Coping with Uncertainty.* Conference Proceedings. Institute for Family and Child Study, College of Human Ecology, Michigan State University: East Langsing, MI.

Morrison, Denton, 1978: Equity impacts of some major energy alternatives *In Energy Policy in the United States: Social and Behavioural Dimensions* ed. S. Warkov New York: Praeger.

Newman, Dorothy and Dawn Day, 1975: *The American Consumer: A Report to the Energy Policy Project of the Ford Foundation.* Cambridge: Ballinger.

Otnes, Per, 1988: Housing consumption: Collective systems service, in Per Otnes (ed.*) The Sociology of Consumption: An anthology.* Solum: Oslo.

Rosa, Eugene, Gary Machlis, and Ken Keating, 1988: Energy. *Annual Review of Sociology.* 14: pp. 149-172.

Schipper, Lee, 1991: Quoted in Cherfas, J. Skeptics and visionaries examine energy savings. *Science* 251: pp. 154-56.

Schipper, Lee, S. Bartlett, D. Hawk, E. Vine, 1989: Linking lifestyles to energy use: a matter of time? *Annual Review of Energy* 14: pp. 273-318.

Schor, Juliet, 1999: *The Overspent American : Upscaling, Downshifting, and the New Consumer.* Harper: New York.

Schwartz Cowan, Ruth, 1983: *More Work for Mother: Ironies of Household Technology from the Open Hearth to the Microwave.* New York: Basic Books.

Schaeffer, G., J., 1998: *Fuel Cells for the Future – A contribution to Technology forecasting from a Technology Dynamics Perspective,* PhD Thesis, Univeristy of Twente.

Shove, Elizabeth and Alan Warde, 1998: Inconspicuous Consumption: the sociology of consumption, lifestyles and the environment, in August Gijswijt, Frederick Buttel, Peter Dickens, Riley Dunlap, Arthur Mol and Gert Spaargaren (eds.) *Sociological Theory and the Environment*, Proceedings of the Second Woudschoten Conference, ISA Research Committee 24, SISWO, University of Amsterdam, pp. 135-154.

Shove, Elizabeth, Loren Lutzenhiser, Simon Guy, Bruce Hackett, and Harold Wilhite, 1998: Energy and Social Systems, in Steve Rayner and Elizabeth Malone (eds.), *Human Choice and Climate Change*, Volume 2, Resources and Technology, Ohio, Battelle Press, 1998. pp. 291-327.

Slater, Don, 1998: Themes from the Sociology of Consumption. ESRC network on *Consumption, Environment and the Social Sciences*, 6-7 July 1998, Oxford.

Spaargaren, G., 1997: The Ecological Modernisation of Production and Consumption: Essays in Environmental Sociology, PhD Thesis, Wageningen Agricultural University.

Socolow, Robert H. (Ed) , 1978: *Saving Energy in the Home: Princeton's Experiments at Twin Rivers.* Ballinger: Cambridge, MA: Ballinger Press.

Socolow, Robert H., 1978: The Twin Rivers program on energy conservation in Housing. *Energy and Buildings* 1(3): pp. 207 –242. Reprinted in Socolow (Ed) 1978.

Sonderegger, Robert C., 1978: Movers and Stayers: The Resident's Contribution to Variation across Houses in Energy Consumption for Space Heating. *Energy and Buildings* 1(1977/1978) pp. 313-324. Reprinted in Socolow (Ed)., 1978.

Stern, Paul C. and E. Aronson (eds.), 1984: *Energy Use: The Human Dimension*. New York: W. H. Freeman Co.

Stern, Paul, Oran Young and Daniel Druckman, (eds.), 1992: *Global Environmental Change: The Human Dimension*. Washington, DC: National Academy Press.

Warde, Alan, E. Shove, and D. Southerton, 1999: Convenience, schedules and sustainability. Reader, Lancaster Summer School on Consumption, Everyday Life and Sustainability, Lancaster University. August.

White, Leslie, 1975: *The Concept of Cultural Systems: A Key to Understanding Tribes and Nations.* New York: Columbia University Press.

Wilhite, Harold and Lutzenhiser, L., 1998: Social loading and sustainable consumption. *Advances in Consumer Research* 26: pp. 281-287.

Wilhite, Harold, H. Nakagami, T. Masuda, Y. Yamaga, H. Haneda, 1996a: A cross-cultural analysis of household energy-use behaviour in Japan and Norway. Energy Policy 24(9): pp. 795-803.

Wilhite, Harold, H. Nakagami, and C. Murakoshi, 1996b: The dynamics of changing Japanese energy consumption patterns and their implications for sustainable consumption. *Proceedings from the ACEEE 1996 Summer Study on Energy Efficiency in Buildings.* American Council for an Energy Efficient Economy, Washington, D. C.

Wilk, Richard R. and Harold Wilhite, 1985: Why Don't People Weatherize their_Homes?: An Ethnographic Solution." *Energy: The International Journal* 10(5): pp. 621-630.

Wirl, Franz, 2000: Lessons from Utility Conservation Programs. *The Energy Journal.* 21: pp. 87-108.

Group Identity, Personal Ethics and Sustainable Development Suggesting New Directions For Social Marketing Research

Johanna Moisander

1. Introduction

Market-based environmental policy programs, associated with climate mitigation for example, usually involve social marketing interventions, in which marketing is exported to the domain of non-profit organizations and public policy as a 'technology' of social change and behavioural influence (e.g., Kotler and Roberto, 1989; Andreasen 1994), and used as an instrument of social control. Such interventions usually take the form of public disclosure and education campaigns, and behaviour modification campaigns, in which basic principles of marketing management are combined with theories from social psychology and consumer theory, in an attempt to design policies and programs to promote sustainable development. The aim of these campaigns is usually to educate and persuade consumers to behave in ecologically responsible ways and, also, to foster collective and business compliance to environmental norms and regulations indirectly through an informed public opinion. Consumers and green consumerism[1] are given a significant role in the transition to a sustainable economy since it is believed that informed citizens and non-governmental and community groups not only assume personal responsibility for environmental protection but also take action, exposing and denouncing private firms and government agencies that abuse the environment (Fuller, 1996; Puranen, 2000; Steer, 1996). Accordingly, then, the research on environmental protection-related social marketing has largely focused on the dynamics of personal ethics and group identity[2], usually labeled as green consumerism, to find ways to influence and appeal to consumers' intrinsic motivation, and to make use of group dynamics and social influence in environmental policy.

The discussion on personal ethics and group identity has largely revolved around intrapersonal psychological processes, particularly motivational, perceptual and informational processes, and individual choice. Owing to the strong position of information processing models in consumer research (Uusitalo and Uusitalo, 1980; Belk, 1987; Hackley and Kitchen, 1998), human behaviour, in these accounts, has

1 In green marketing and social marketing literature discussing environmental policy the tern 'green consumerism' usually refers to consumer behaviour that reflects and exhibits, in various ways, a concern for the environment (e.g., Henion, 1974). In green marketing text books (e.g., Ottman, 1998; Peattie 1992), for example, the green trend in consumer values and attitudes is usually viewed as a new market potential for environmentally safe products and services, and also as a threat, in the form of consumer boycotts and decreased market shares as consumers "vote with the dollar" against companies which neglect the management of the environmental impacts of their business activities.

2 Group identity, in social marketing literature, is usually conceived of as a person's identification with a social group with distinct symbols and characteristics. It is generally also assumed that a particular group identity, such as a conception of self as a green consumer, involves a set of behavioural or personal characteristics by which an individual is recognizable as a member of that group.

E. Jochem et al. (eds.), Society, Behaviour, and Climate Change Mitigation, 127–156.

generally been viewed as largely governed by a disembodied 'mind' (Joy and Venkatesh, 1994; Peñaloza, 1994), whose interpretation of his/her experiences and the external, social world has been studied with latent mental constructs, such as attitudes and values, usually inferred from observed or measured responses and behaviour. Moreover, in line with the microeconomic underpinnings of both marketing theory and market-based environmental policy, environmental protection-related social marketing problems have usually been framed in terms of individual choice and markets. As Andreasen (1991) has put the point, the question being asked is: how do we make the marketplace work better so that consumers can make better decisions about what to buy? Accordingly, consumers have traditionally been conceptualized as fairly autonomous and reasonably rational decision-makers who act on beliefs, express attitudes, pursue personal values, strive toward goals, and make more or less informed choices and purchases among available alternatives (e.g., Kalafatis *et al.,* 1999; Pieters *et al.*, 1998; Pelton *et al.,* 1993; Thørgesen, 1994; Taylor and Todd, 1995).

In more elaborate analysis, the social and behavioural dynamics of environmentally responsible behaviour have been theorized and explained, more or less explicitly, in terms of strategic choice behaviour involving a social dilemma in producing public or collective goods, or a tragedy-of-the commons situation[3] (Berger and Kanetkar, 1995; Fuller, 1999; Gardner and Stern, 1996; Hutton and Ahtola, 1991; Pieters, 1991; Uusitalo, 1989; Wiener and Doescher, 1991; Wiener, 1993). Still, environmental problems have been explained with self-interested consumers behaving in a manner that maximizes their individual utility by taking the role of free riders in the production of environmental quality.

Therefore, in the contemporary social marketing literature green consumerism is usually conceptualized as individual, pro-social or ethically oriented consumer behaviour. Accordingly, then, various persuasion strategies appealing to consumers' personal environment related moral norms, their group identity and their important role as socially responsible green consumers, are frequently advocated – in addition to financial incentives – to promote cooperation in favor of environmental quality. Consumers' self-interest, feelings of powerlessness and insignificance in protecting the environment, as well as their mistrust in others' cooperation are viewed as major barriers to environmentally responsible behaviour and, therefore, it is suggested that these beliefs and feelings be targeted in environmental protection-related social marketing programs (Dawes *et al.*, 1990; Fuller *et al.*, 1996; Granzin and Olsen, 1991; Ellen *et al.*,1991; Uusitalo, 1989; Wiener and Doescher, 1991; Wiener, 1993).

In much of the literature focusing on these issues, it is assumed that cooperation in environmental protection can be promoted through some sort of change in people's

---

3 In tragedy-of-the-commons situation, behaviour that makes sense from the perspective of a self-interested individual ultimately proves disastrous to society (Hardin, 1968). In the context of environmental protection, this problem arises from the fact that environmental quality is to a significant degree a collective or public-type good, which is produced and consumed collectively, has no market value (price), and to which all members of the collectivity have unrestricted access independently of the size of their personal contribution to the production of the good. (Turner *et al.*, 1994; Uusitalo, 1989, 1991).

beliefs and attitudes invoked by correct information. It is believed, as Fuller et al. (1996:68) has put it, that "expectations for rational behaviour depend on successfully translating, packaging, and disseminating information about the issues". Thus, ignorant and incapable consumers should be informed and educated to adopt ecologically correct ways of conduct (Pieters, 1991; Thøgersen, 1994). Mistrustful, skeptic and helpless consumers should be provided with feedback and information about progress in environmental protection, and about the participation of other people, firms and governmental offices in environmental policy programs (Uusitalo, 1989; Ellen *et al.,* 1991; Wiener and Doescher, 1991; Wiener, 1993), and self-interested consumers should be persuaded with appeals to intrinsic motivation and group solidarity (Carman, 1992; Pelton *et al.,*1993). The existing research on green consumers has, thus, tended to view awareness and knowledge as the prime mover of consumers, and something logically and temporally prior to affective reaction (Hackely and Kitchen, 1999) in environmentally responsible consumer behaviour.

It is argued in this paper, however, that the dominant individualistic and choice-theoretic approach to studying green consumerism, and the techno-economic policy implications that it produces, seem insufficient and inadequate in some respects for delivering the potentially valuable contribution that social scientific research might have for sustainable development. It is argued that in many important ways green consumerism is a social and cultural phenomenon. It involves complex sociocultural and socioeconomic relationships in socially and culturally specific environments that "go beyond the determinist Newtonian-type cause-effect or stimuli-response models" (Kumcu, 1987:117) typically found in traditional, marketing-related consumer research (Firat *et al.,* 1987; Ger, 1999; Heiskanen and Pantzar, 1997; Meriläinen *et al.,* 2000). Therefore, the dominant approach to studying green consumerism seems very limited in capturing the full richness and social embeddedness of human experience and behaviour (Hackely, 1998; Potter 1996; Potter and Wetherell , 1987; Wetherell, 1996). It may be argued that, for the most part, the findings that this type of research is currently producing are rather trivial. As Kenneth Gergen (1996) has put the point, from a constructionist perspective the traditional attempt to test hypotheses about universal processes of the mind (e.g., cognition, attitudes, self-conception), based on a uniquely Western ontology, seems misguided, at a minimum, and perhaps even an enormous waste of intellectual, monetary, and time resources.

Consequently, it is the thesis of this paper that one needs to be epistemologically reflexive when using consumer inquiry to design environmental protection-related social marketing interventions. Thus this paper aims to reflect critically on the kind of understanding or knowledge that environmental policy oriented social marketing research produces of the role of group identity and personal ethics in green consumerism. In the sections that follow, I tend to question the hegemony of the conceptual frameworks, methods and styles of research adopted from natural sciences, via microeconomics and mainstream social psychology, which largely postulate that the subject matter of social marketing is social behaviour that can be observed and measured, and which tend to objectify (Gergen 1996) or at least emphasize (Uusitalo 1991) or provide implicit support for individualist ideology and institutions. In social marketing-related consumer research, accounts of environmentally responsible behaviour have been faulted for placing too much emphasis on individual choice and for neglecting the social construction of pro-environmental attitudes, personal environmental ethics, and the identity labeled as 'green consumer'. I argue that more attention should be paid to the representation of green consumerism and the culturally shared moral ideas, meanings, and signifying practices that are typically associated with green consumerism.

The paper is structured as follows. First, the conventional way of conceptualizing and theorizing group identity and personal ethics in social marketing-related consumer behaviour research is presented. Then the conceptualization of 'green consumer' as a group identity and the approach to moral behaviour in recent studies on environmentally responsible consumption are discussed. Finally, conclusions are drawn and suggestions for future research and social marketing programs are given.

2. Experimentalist approach to social behaviour: group identity and personal ethics as intrapersonal variables

In an attempt to construct widely generalizable or universal explanations, applicable to all purposes, conditions, and situations, consumer researchers focusing on green issues have traditionally studied the role of personal ethics and group identity in environmentally responsible consumption as intrapersonal causal determinants of behavioural choice. Owing to its disciplinary background and theoretical basis in social psychology, cognitive psychology and neo-classical microeconomics, consumer research usually takes an *experimentalist*[4] perspective to studying human beings, which Lalljee (1996:93) has characterized as "neither positivist nor behaviourist, while having been influenced by both". Belk (1987) has provocatively described this conception of the human being as an elegant synthesis of 'rational economic man' postulated in neo-classical economics and cognitive psychology's metaphorical model of man as a computer. Like neo-classical microeconomics and traditional experimental social psychology, this style of research, typical of most questionnaire studies and laboratory experiments in mainstream marketing and social psychology, assumes an individualistic conception of subjectivity or agency (Gergen, 1996; Nelson, 1997), and draws its general orientation from natural sciences. It thus presumes a tangible reality with discrete elements and causal relationships, and emphasizes measurement, reliability and objectivity.

Consumer research on green consumerism has primarily been focused on psychological, intrapersonal phenomena, such as attributions, inferences, judgments, and decisions. To give an example, the studies focusing on environmentally responsible behaviour that were published in *Journal of Public Policy and Marketing* and *Jour-*

---

4 The term 'experimentalist perspective' is currently used to refer to the dominant "experimental paradigm" in social psychology (e.g., Gergen, 1996; Stevens, 1996). This perspective is usually seen as a "conservative", "positivist", or "modernist" approach to studying social behaviour, because, it has tended to (1) place a strong emphasis on intrapersonal psychological states and processes, (2) presume the pre-discursive reality of the subject matter of research, (3) rely on some form of foundationalism, (4) presume a correspondence view of language, and see the scientist as an informant to the culture, and (5) treat scientific/scholarly effort as politically/ideologically neutral (Gergen 1997).

nal of Macromarketing in the '90s, were all based on such experimentalist designs. All of them were using questionnaire studies or laboratory experiments (typically with student samples) usually to test narrowly defined hypotheses about the effect of intrapersonal variables on judgment or behaviour (see Appendix 1).

In general, this research appears to have been informed by cognitively oriented social psychology, assuming that to understand human behaviour and social experiences it is necessary to understand the "precise nature of the [internal] representations" (Markus and Zajonc, 1985: 138), and the cognitive structures that individuals use to recognize and interpret information and social stimuli from the external environment. Motivational factors, such as attitudes, values and moral ideals are thus viewed as characteristics of individuals, and even the social and cultural dynamics of green consumerism are often studied in terms of personal mental constructs, such as 'subjective norms' (e.g., Taylor and Todd, 1995; Kalafatis *et al.,*1999). The "individualist subject" (Hall, 1992) that this type of research implies thus appears to be constituted by its capacity to reason, think, and act. Moreover, as Gergen (1996) has observed, this traditional approach to human subjectivity, tracing human action to psychological factors, sustains a view of people as fundamentally isolated, self-gratifying, and self-sufficient.

Owing to its epistemological and methodological commitments, social marketing-related consumer theory has, nevertheless, customarily studied these intrapersonal phenomena from the outside, reducing a person's subjective experience and complex psychosocial phenomena to fairly simple responses that can be observed and quantified with measures of verbal response, categorization, recall, recognition, and reaction time, for example. In doing this, researchers have generally accepted that abstract social and psychological phenomena can be broken down into components and treated almost as if they had a concrete or material existence. Accordingly, a person's social identity and moral values have been studied mainly by isolating and reducing these concepts to a number of simple variables, accessible to direct measurement or manipulation, that can be used to formulate empirically verifiable statements. Ethically oriented behaviour, for example, has typically been conceptualized as a pre-defined personal norm or a "perceived moral obligation" (e.g., Osterhus, 1997), which has been operadionalized as internalized moral rules, such as "I feel obligation to save energy where possible" (ibid.), and measured with structured questionnaires. Knowledge of people's verbally expressed agreement with such statements, however, would appear to capture only partly their personal beliefs about what is right and wrong in the context of environmental protection (as will be discussed in the next section of this paper).

The typical epistemological and ontological commitments of experimentalist research appear to reflect the idea that since humans belong to the natural order, a single broadly defined method will serve for all sciences (Hollis, 1994). It may be argued, however, that as the social world is constructed from within, in a way quite alien to the natural world, social sciences must rely on inter-subjectivity and seek the meaning of action. Behaviours, talk, texts and images derive their meaning from the shared ideas and rules of social life, and are performed by actors whose everyday life is characterized by a complex weave of interests and entitlements (Hall, 1997,

Stevens, 1996; Potter, 1996; Wetherell, 1996). Therefore, it would seem fruitful not only to generate 'objective' experimentalist, logical-empiricist knowledge about our social behaviour but also to produce knowledge about what is socially constructed as true, and perhaps also experientially real, rather than abstractly true (Stevens 1996). Such knowledge would seem helpful for stimulating awareness and reflection on the roles and responsibilities of researchers, consumers, firms, and other societal actors in sustainable development.

3. Group identity and personal ethics as motives for environmentally responsible consumption

In social marketing research, personal ethics and group identity have usually been viewed as significant motivators of environmentally responsible consumer behaviour (Granzin and Olsen, 1991; Moisander, 1996; Osterhus, 1997; Sparks and Shepherd, 1992; Wiener, 1993; Uusitalo, 1989, 1991). Green consumerism has traditionally been viewed as a specific type of *socially conscious* (Anderson and Cunningham, 1972; Webster, 1975) or *socially responsible* (Antil, 1984; Pelton *et al.*, 1993; Singhapakadi and LaTour, 1991) consumer behaviour, which is postulated to be motivated not only by consumers' own personal needs but also their considerations of the welfare of society in general (Antil and Bennet, 1979). In line with the expectancy-value approach to human motivation (e.g., Fishbein and Ajzen, 1975; Ajzen and Fishbein, 1980) that has been dominant in traditional marketing and consumer research, the socially conscious consumer is viewed as a goal-oriented decision-maker who

> *takes into account the public consequences of his or her private consumption or ... attempts to use his or her purchasing power to bring about social change,*

as e.g., Webster (1975:188) has defined the concept. Thus ecologically oriented consumer behaviour is generally presumed to be motivated by two different types of consumption goals: the individual objectives of the consumer and the collective objectives of society, i.e., environmental protection and sustainable development (e.g., Pickett *et al.*, 1993; Singhapakdi and LaTour, 1991). Moreover, ecologically responsible consumer behaviour is customarily framed in terms of individual choice based on evaluations of both personal and social costs and benefits associated with different behavioural or product alternatives (e.g., Pieters, 1991; Powers *et al.*, 1992; Taylor and Todd, 1995). Little, if any, explicit thought or empirical evidence, however, is usually given to substantiating the assumption that consumers are actually goal-oriented and think in terms of balancing personal costs and benefits. Apparently drawing from the neo-classical metaphor of 'rational economic man', ecologically responsible consumers are assumed to be motivated by considerations of private and public utility.

In line with the microeconomic foundations of social marketing theory, these choice situations are generally seen as involving a type of social dilemma or a many-party case of the prisoners dilemma, or a tragedy-of-the commons situation (Berger and Kanetkar, 1995; Fuller, 1999; Gardner and Stern, 1996; Hutton and Ahtola, 1991;

Pieters, 1991; Uusitalo, 1989; Wiener and Doescher, 1991; Wiener, 1993). In a social dilemma situation, as described by Dawes (1980: 170),

> *...the social payoff to each individual consumer for defecting behaviour, is higher than the payoff for cooperative behaviour, regardless of what other society members do, yet, ...all individuals in society receive a lower payoff if all defect rather than cooperate.*

Ecologically responsible consumption is, thus, viewed as involving a collective-individual paradox; owing to the fact that environmental quality is a collective good, i.e., free for everyone but produced collectively, what is rational for the individual is not necessarily rational collectively (Uusitalo 1991). The contribution of a single consumer to environmental quality is so small that it may be tempting, and would seem rational from an individual consumer's point of view, to take the role of a free rider and, without own effort, benefit from others' efforts in producing the environmental quality. Yet, the more consumers contribute to the environmental quality the better it is and, therefore, it is in the best interests of all ecologically responsible consumers that everybody in society cooperates. Or to put it more precisely, in consumers' preference ranking, the situation that everyone cooperates is higher than the situation that no one cooperates. Therefore, to be environmentally responsible, according to this line of thinking, consumers have to engage in socially responsible or altruistic behaviours that are in conflict with their private utility and with principles of individual rationality. Consequently, ecologically responsible consumer behaviour is seen to involve a *commitment problem* (Frank, 1988): willing to be ecologically responsible, the consumer wants to cooperate, i.e., contribute to the collective production of environmental quality, but the short term benefits from defecting, i.e., free riding, are tempting.

The solutions to the social dilemmas and the associated commitment problems involved in ecologically responsible consumer behaviour, generally proposed in social marketing literature, are usually based on suggestions for removing psychological 'barriers' to cooperation. Wiener (1993), based on Wiener and Doecher (1991), for example, has identified three obstacles that have been widely adopted in social marketing research:

- self-interest, a temptation to free ride or a perception that the social payoff is too small compared with personal sacrifice;
- mistrust or suspicion that others are not cooperating; and
- a fear of being a "sucker", i.e., a fear that the goal of sustainable development will not be achieved even if the sacrifice associated with cooperation is made.

The proposed solutions to these problems associated with social dilemmas are basically of two types: solutions based on self-interested motives and those based on unselfish motives. The self-interested solutions usually refer to various side payments, selective financial incentives, social sanctions, punishments, laws and regulations and the like, which change the pay-offs in a social dilemma situation, adding something extra to the pay-offs of anyone who cooperates. The unselfish solutions to the commitment problem associated with ecologically responsible consumer behaviour refer to the integration of the long term collective utility into the personal utility function of the consumer (Frank, 1988) or to "selfless", "purely charitable" altruistic motives (Singhapakdi and La Tour, 1991).

In social marketing literature, suggestions for overcoming these psychological barriers to cooperation usually emphasize the role of group identity and personal ethics in motivating ecologically responsible behaviour. On the basis of classic psychological theories of social reference and group dynamics, identification with a reference group of green consumers is believed to enhance cooperation through shared group norms, values, and information. Group identity and the feelings of we-ness, i.e., "we are in this together", and sharing a common fate – or facing a common threat – involved in social dilemma situations (Granzin and Olsen, 1991) have been found to help people in overcoming suspicion, skepticism, and the feelings of, insignificance and powerlessness (Granzin and Olsen, 1991; Wiener, 1993; Wiener and Doescher, 1991). Dawes et al. (1988) have also reported findings suggesting that group solidarity increases cooperation independently of side payments, both internal and external, often associated with group identity.

Personal proenvironmental ethics with internalized moral norms have been proposed as a solution for overcoming the psychological barrier of self-interest in the adoption of ecologically responsible ways of life (Uusitalo, 1989, 1991; Wiener and Doecher, 1991; Wiener, 1993). Moral beliefs and the specific emotions that they are associated with are seen to function as some sort of psychological incentives for cooperative behaviour. Social marketing-related literature largely lacks a more detailed discussion on the dynamics of personal ethics in green consumerism, but some scholars have suggested that internalized moral ideals involve a redefinition of self by incorporating some external moral ideals (e.g., equity, fairness, reciprocity, justice) into self-identity (Jencks, 1979). These ideas are apparently grounded on the assumption, dominant in cognitively oriented social psychology, that a person's self-concept mediates and regulates his or her social behaviour. As Markus and Wurf (1987, 299-300) have pointed out, 'self', in modern self theory, is viewed as a dynamic mental construct that interprets and organizes self-relevant actions and experiences; *has motivational consequences, providing incentives, standards, plans, rules, and scripts for behaviour;* and adjusts in response to challenges from the social environment. Hence, internalized environmental protection related moral norms and ideals are viewed as causing the individual to behave 'unselfishly', or to cooperate in a social dilemma situation, because failure to follow these ethical principles produces an inner conflict of some sort, producing guilt and bad feelings (Frank, 1988). In other words, consumers who have internalized some moral norms associated with environmental protection are thought to feel obliged to behave in an ecologically responsible way regardless, to a degree, of personal sacrifices that are needed (e.g., Granzin and Olsen, 1991; Osterhus, 1997).

Social marketing research focusing on green consumerism as a case of socially responsible consumer behaviour involving a social dilemma has undoubtedly produced an interesting body of knowledge on the dynamics of strategic choice behaviour. However, further elaboration of the observed – usually weak – correlations and hypothetically causal relationships with improved measures and more sophisticated mathematical methods and models seems to produce only marginal contributions to our current understanding of green consumption. A major problem with this line of theorizing resides in the conceptualization of green consumerism in terms of individual choice, guided by personal ethics, social norms and a self-conception of a 'green consumer' – understood as intrapersonal motives or psychosocial incentives for cooperative behaviour in tragedy-of-the-commons type consumption situations. This 'individualist' conception of the subject, characterized by independent thought and rational (moral) reasoning originating from universal and biological processes of mental functioning (Gergen 1996), seems fictitious. As Gergen (1996) has put the point, such a view of human beings perpetuates the long-standing individualist practices within the culture, stressing the independent functioning of the individual and relegating social institutions to mere by-products of individual interaction. Thus such an approach fails to pay attention to, or investigate, how consumers perceive and represent their roles and responsibilities in sustainable development and how moral considerations and group identity as a 'green consumer' are constructed, expressed and acted out in consumption.

4. 'Green consumer' as a group identity

In most social marketing texts, the identity of the green consumer is understood as being grounded on ecologically oriented consumption activities that exhibit and reflect a relatively consistent and conscious concern for the environment (Henion, 1976). However, the particular accounts of this 'environmental concern' and, thus, of the nature of group identity associated with green consumerism seem to vary considerably in recent literature.

4.1 CONCEPTUALIZATION OF THE GREEN CONSUMER

The specific ways that green consumers and green consumerism have been conceptualized in environmental protection-related social marketing literature are many. Sometimes a green consumer is defined simply as "anyone, whose purchase behaviour is influenced by environmental concerns" (Shrum *et al.,* 1995: 72). Sometimes a green consumer is seen to have adopted a "lifestyle that has minimal adverse effects on the biophysical environment" (Banerjee *et al.,* 1995, 22). Being green, in these accounts, is viewed as involving an assessment of the environmental impact of product/service choices, and a behavioural change in purchasing, consuming, and disposing of the product (e.g., Banerjee *et al.,* 1995; Henion, 1976). Moreover, in the green marketing literature, the environmentally concerned consumer is also increasingly associated with environmentalism and viewed as a person who is actively involved in environmental or green consumerism (Elkington *et al.*, 1990; Ottman, 1998). Green consumers are seen as significant market actors who, having teamed up with environmental organizations and pressure groups, such as Greenpeace, demand the business world should clean up its act and take responsibility in sustainable development.

Except for some segmentation studies (e.g., Pilling *et al.,* 1991), being green is typically understood and conceptualized as a one-dimensional concept or characteristic, essentially a continuous variable ranging from 'very green' to 'not green at all' (e.g., Banerjee *et al.,* 1995; Fuller *et al.,* 1996; Ottman 1998; Vlosky *et al.,* 1999). Accordingly, using behavioural and demographic segmentation criteria, consumer markets are customarily divided into a set of progressively greening segments ranging, for example, from extremely committed "True Blue Greens" to "Basic Browns" who express the least amount of environmental concern and involvement in their behaviour (Roper Starch Worldwide Inc., 1996, ref. in Fuller, 1999).

In general, the 'more green' consumers are usually defined as people whose purchase is greatly or very likely to be influenced by environmental concern; and 'less green' as those whose purchase behaviour is only minimally or unlikely to be influenced by environmental concerns (e.g., Shrum, 1995: 72; Vlosky *et al.,* 1999). More detailed accounts of what this concern means tend to differ. Sometimes distinctions are made on the basis of observed variation in the frequency with which people perform certain simple behaviours, which are taken as instances of a more general behavioural category of 'environmentally concerned behaviour'. Sometimes consumers' greenness is quantified by pre-labeling certain behaviours as expressing little commitment to 'greenness' and other behaviours as indicating more serious commitment to a green lifestyle (e.g., Ellen *et al.,* 1991). Some scholars acknowledge that green consumers may differ on trait and attitudinal variables that are more specific than the broad demographic and simple behavioural variables but, nevertheless, make the distinctions between consumers primarily in terms of what they appear to perceive as quantitative differences in consumers' 'greenness' (e.g., Shrum *et al.,* 1995).

It is argued here, however, environmental concern is a multidimensional concept and green consumerism takes many forms among ecologically oriented consumers (Moisander, 1996). Moreover, as Darier (1999) points out, because there are no absolute external referential categories for greenness, the degree of 'greenness' can only be measured in context, not in the abstract. Therefore, it does not appear to be very useful to distinguish between more and less green consumers and to consider 'green consumerism' as essentially a one and the same group identity.

4.2. MYRIAD WAYS OF THINKING, BEING, AND ACTING GREEN

The ways consumers express their identity as 'green consumers' in their everyday lives are multitudinous, and there is hardly an agreement upon the appropriate means and ends for 'green consumerism'. Even among the 'True-Blue Green' consumers there may well be considerable disagreement upon how the concern for environmental consequences of consumption activities is or should be manifested in consumer behaviour.

Firstly, there may be *divergent views on the basic objectives and strategies of ecologically responsible consumption.* As Elkington et al. (1990: 5) point out, one view is to hold that to truly care for the environment means drastically reducing the number of purchases of everything to a bare minimum. An alternative view is to acknowledge that such a radical environmentalist approach to consumption is not easy to adopt in our increasingly convenience- and consumption-oriented society. According to this more liberal view on green consumerism, by carefully choosing products and services that are the least destructive to the environment it may be possible to have a positive impact on the environment without significantly compromising one's way of life.

Secondly, the difficulty of defining environmentally sound consumption strategies is further increased by the fact that there are *no agreed-upon criteria for what constitutes an ecologically sound or safe product or service.* Some general properties of green products and services may, however, be depicted. Elkington et al. (1990: 6), propose the following criteria. Environmentally sound products and services:

- are not dangerous to the health of people or animals
- do not cause damage to the environment during manufacture, use, or disposal
- do not consume a disproportionate amount of energy and other resources during manufacture, use, or disposal
- do not cause unnecessary waste due either to excessive packaging or to a short life span
- do not involve the unnecessary use of or cruelty to animals, and
- do not use materials derived from threatened species or environments.

Nevertheless, these criteria remain – as Elkington et al. (1990: 7) quite readily admit – somewhat obscure because people's perceptions of these issues may be very different. For example, it is impossible to objectively define what constitutes 'unnecessary waste' or a 'disproportionate amount of energy' because assessments of these qualities involve personal value judgments and, thus, vary from one person to another.

Thirdly, there may be disagreement upon what are the *relevant elements of environmental quality*, i.e., which are the relevant areas of environmental concern, and which elements should be given priority in consumers' ecologically responsible consumption strategies. The areas of environmental concern are many and deal with various environmental protection issues ranging from conservation of natural resources and urban waste problems to human population issues (e.g., Scheffer 1991). Moreover, some researchers also include cultural and aesthetic elements into the concept of environmental concern (e.g., Uusitalo 1986).

Fourthly, people's conceptions of ecologically responsible consumption may vary in terms of *what are the relevant behavioural elements involved*, i.e., which specific behaviours are considered ecologically relevant, as well as in terms of *what is the weight or magnitude of each behaviour* involved in their patterns of ecologically responsible consumption.

The behavioural elements or the sets of behaviours people include in their personal ecological consumption strategies may be very different because people seldom act out their environmental concern in each and every relevant aspect of behaviour. Few ecologically minded consumers decide to do everything right, or in an environmentally responsible manner. More probably, the majority of green consumers do only what they perceive as their fair share of the things that they know and come to think of as environment friendly forms of behaviour that can be done (Bell, 1994; Kempton, 1991; Moisander, 1996). Nonetheless, although people do not regularly engage in some or many of the ecologically relevant behaviours they know of, they may still consider themselves as green consumers.

Furthermore, the sets of behaviours consumers include in their ecological consumption patterns may also vary in terms of the extent to which each chosen, ecologically sound behaviour is performed. Some consumers may, for example, be unwilling or unable to use public transportation to commute but they engage in other energy saving behaviours extensively, thus trying to compensate for their environmentally harmful commuting behaviour. Some other consumers, on the other hand, may decide to give up driving to work for environmental reasons but they do not feel the need to engage in some other energy saving forms of behaviour.

And finally, besides owing to the great variability in setting goals and in acting out green consumerism, it does not appear to be very useful to distinguish between more and less green consumers or to consider 'green consumerism' as one and the same group identity because *different people may be 'green' for different reasons.* For some, the representation of green consumerism may primarily be associated with acting out their social role as parents and is thus based significantly on a concern for family security (Moisander 1991, 1997). Some others may tend to be more oriented towards personal growth and physical wellbeing in representing themselves as green consumers (Fjeld *et al.,* 1984; Moisander, 1991). For yet another group of people, it may well be a set of certain ideological, political or religious beliefs, meanings, feelings and values that is constitutive or characteristic of their group identity as green consumers (Milbrath, 1986; Moisander, 1997).

Furthermore, the complex sets of meanings, values, and ideas that people's personal orientations to green consumerism involve may, of course, change over time, taking different forms in different social contexts. Considering green consumerism as a fixed, closed and unchanging group identity seems, therefore, unwarranted (Gilroy, 1997; Hall, 1996). Being green may arguably be experienced differently depending on one's age, class, ethnicity, gender, and possibly also on a number of other personal factors. Consequently, to get a more profound understanding of the social and behavioural dynamics of green consumption, it would seem important to recognize explicitly that the role of 'green consumer' intersects with other social identities, and the associated interests and entitlements, during the practice of daily life.

Moreover, considering the socio-political and global economic complexity of the phenomenon, it might be illuminating to look at green consumerism also as a predominantly social movement, emphasizing its role as a countervailing power against the business world. Antonides and van Raaij (1998:102) have distinguished several types of general consumerism based on somewhat different underlying political and societal commitments:

- Liberal consumerism, focusing on looking after the rights and increasing the power of consumers in the markets;
- Responsible consumerism, focusing on the societal responsibility of consumers;
- Critical consumerism, focusing on questioning some features of the prevailing consumption system, e.g. demanding legislation and government action for better and safer products; and
- Radical consumerism, focusing on attacking the capitalist ideology of consumption, e.g. the inequity of the societal system that it involves.

As a social movement green consumerism has arguably taken shape in ways that bear a resemblance to many of the characteristics attributed to these general orientations to consumerism. Therefore, it would seem useful to investigate green consumerism as a social and cultural phenomenon also from the social marketing point of view. For a deeper and more thorough understanding of the psychosocial 'barriers of cooperation' it would seem important to account for the culturally shared ideas, meanings and signifying practices that are typically associated with green consumerism as a social movement.

5. Morality, rationality and green consumerism

In social marketing-related literature, personal ethics is usually viewed as a system of ethical principles concerning personal conduct that involves conflicts of interests (Fitzgerald and Corey, 1992). It is often viewed as involving a moral imperative, which is seen as a cognitive structure for determining how conflicts in human interests are to be settled, and for optimizing mutual or societal benefits of people living together in groups (Pelton *et al.,* 1993: 64). Moreover, perhaps because of the legacy of the traditional, enlightenment inspired view of human beings, according to which rationality is considered to secure morality (e.g.,Thompson 1993), personal ethics is generally conceptualized as an 'ethical choice' based on morally well reasoned justification for the choice of action, usually based on utilitarian considerations (e.g., Fitzgerald and Corey, 1992; Osterhus, 1997). Thus it is presumed that that green consumers are fairly 'rational', centered actors[5].

---

5 The conception of identity underlying most conceptualizations of green consumerism in social marketing literature reflect what Hall (1992:275), among others, refers to as "the Enlightenment subject". This highly individualist conception of the subject is based on "a conception of the human person as a fully centered, unified individual, endowed with the capacities of reason, consciousness, and action". This autonomous subject, which is often also referred to as the Cartesian subject, is viewed as having some kind of stable, essential, uniquely individual, and pre-discursive core or center, or a fixed, well recognized 'self', which is capable of independent and rational thought. According to this line of thinking, human beings are free to the extent that their actions are carried out for a reason, and human rationality is universal, requiring only education for its development (The Cambridge Dictionary of Philosophy 1999:266).

5.1. PERSONAL ETHICS AND GREEN CONSUMERISM

In social marketing literature the role of personal ethics in green consumerism is usually studied using a normative-descriptive approach (Fitzgerald and Corey, 1992) to morality. Researchers typically rely on a normative approach to ethics when constructing their theoretical frameworks for investigating moral dimensions of green consumerism, and then employ a descriptive approach to ethics in reporting the ethical judgements of subjects. Recurrently, on the basis of some largely unelaborated and taken-for-granted ideas of what represents a moral action in a given decision situation, researchers postulate that 'morally oriented consumers' choose or behave in a particular way predefined as 'moral' or as the 'right' thing to do (e.g., Fitzgerald and Corey, 1992; Osterhus, 1997; Sparks and Shepherd, 1992). It is first presumed, for example, that moral behaviour demonstrates a concern for others and their wellbeing, and then it is studied empirically whether or not subjects express this concern in behaviour or judgment. Little attention is paid to the nature of moral reasoning and the conceptions of morality of the subjects themselves.

In line with the strong faith in instrumental rationality in traditional scientific thinking, studies focusing on green consumerism as ethical consumer behaviour have often assumed a teleological and anthropocentric view to personal ethics (e.g., Granzin and Olsen, 1991; Pelton *et al.,* 1993; Pieters, 1991; Singhapakadi and LaTour, 1991; Wiener, 1993). In attitude research, for example, moral obligation is often subsumed under attitudes because guilt, self-reinforcement, and other outcomes of meeting or violating one's own standards are seen merely as additional consequences of behaviours (e.g., Ajzen and Fishbein, 1980). When the influence of personal ethics on consumer behaviour is explicitly investigated, consumers are generally viewed, usually implicitly, to ground their moral considerations in some sort of teleological reasoning about human interests. As environment-related moral decisions are usually conceptualized as involving a fundamental conflict between individual and collective utility, or personal and social benefits, a given decision is usually considered 'moral' and 'right' if it brings about positive consequences for all people involved, preferably producing the greatest good for the greatest number of people. Pelton et al. (1993), for example, conceptualize environmentally responsible behaviour as ethical decision-making based on weighting egoistic and utilitarian decision outcomes. Thus green consumerism is considered moral because it has positive consequences not only for the actor but also for all people (present and future generations of human beings) affected by consumption activities.

Significantly less attention has been paid to de-ontological moral considerations underlying green consumers' ethical judgments and behaviours. The idea that people would engage in rule based moral reasoning, judging their actions as inherently right or wrong, independently of their anticipated consequences, is sometimes implied in theories discussing altruism (e.g., Schweper and Cornwell, 1991), but little attention has been paid to how such moral obligations are constructed. There seem to be grounds for assuming, however, that being a deeply concerned and dedicated green consumer often involves a commitment to some sort of a prima facie duty, i.e., a duty that is obvious or evident without proof or reasoning. Harré et al. (1999), for example, have found that the moral discourse that many environmentalist organizations such as Sierra Club rely on is de-ontological in flavor, placing an emphasis on doing something right in itself rather than for some practical end.

Moreover, the extant research on green consumerism has largely presupposed unreflectively that green consumers prescribe to some sort of anthropocentric approaches to environmental ethics. Few explicitly ecocentric accounts have been proposed. Yet, some deeply ecologist consumers might well believe that the flourishing of both human and non-human life on Earth has intrinsic value and that the value and rights of non-human life forms are independent of the usefulness they have for narrow human purposes (Naess, 1989, ref. Mappes and Zembaty, 1992: 479). Some might argue that sustainable development would downright require that consumers ascribe some kind of objective value to nature and adopt a moral obligation to protect it. It may well be that some green consumers consider a given environmentally sensitive moral decision 'right' primarily because of some religious moral beliefs or moral obligations based on deontological and/or ecocentric ethical considerations, such as the respect for 'mother earth'.

In sum, a teleological and anthropocentric approach to personal ethics is customarily presupposed in environmental protection-related social marketing research focusing on green consumerism. In the case of experimentalist research designs, where subjects are asked or made to respond to a set of pre-structured stimuli based on a preconstructed research problem, such an approach to studying environmental ethics seems of severely limited use for improving our understanding the moral dimensions of ecologically responsible behaviour. In practice, the dominant normative-descriptive approach to studying personal ethics would seem to mean that the moral rules and evaluative standards that are pre-defined as constitutive of a person's system of moral norms are postulated as valid for all green consumers in all contexts. In other words, a particular system of ethics is in fact imposed on the subjects of the study.

5.2. THE 'RATIONALITY' OF ENVIRONMENTALLY MORAL BEHAVIOUR

On the whole, it would seem that the individualist view of human being as a centered rational subject has guided research on green consumerism to view people as having stable preferences and instrumentally motivated moral norms that apply to all consumption contexts and situations. But perhaps human beings are not as conscious, rational and principled as we would like to think. Harré et al. (1999), for example, have observed that environmental advocacy represents a certain way of thinking and acting that is essentially a style of life, based on a complex interplay of both moral an aesthetic criteria, which on close examination are not completely stable.

A constantly growing number of social scientists would seem to agree that people are not rational in any conscious, stable, continuous and de-contextualized manner (e.g., Gergen and Joseph, 1997; Firat and Venkatesh, 1995; Hall, 1996; Prior, 1997; Wetherell, 1996). At the extreme, the concept of individual rationality is found not only conceptually flawed but also oppressive in implication (Gergen and Joseph,

1997). As Elisabeth Grosz (1990) has argued, already early in the beginning of the 20th century, Nietzche, Marx and Freud challenged the prevailing Cartesian conception of human subjectivity, decentering the individual's pretension to sovereignty, self-knowledge, and self-mastery (also Hall, 1992). Taking consciousness as the consequence of corporeality, Nietzsche viewed reason not so much of a quality or attribute of the mind as the result of political and coercive struggles between various competing perspectives, in which one gains a provisional, temporary, historical dominance. Marx, on the other hand, pointed out the ideological, misled and untrustworthy, 'false consciousness' constituted by a structure of class relations. Freud, in turn, viewed consciousness and its self-certainty as the end-product of unconscious psychical 'defences' (denial, disawoval, resistance) and, thus, based to some extent on self-deception.

Particularly Freud, according to Grosz (1990: 2), with his understanding of the unconscious, sexuality, psychical representations, and the processes involved in the constitution of the subject, has challenged the Cartesian subject's status as the foundation and source of knowledge. Grosz (1990: 2) argues:

> *If the subject is necessarily incapable of knowing itself – that is, if there is unconscious – then its claims to found knowledge of the world on the certainty of its own existence are also problematized. If the subject cannot know itself, why should we believe it can know anything else with absolute certainty? This does not entail scepticism or nihilism, whereby knowledge is impossible, but it may imply that knowledge, consciousness, and subjectivity need to be reconceptualized in different terms and assessed by new criteria not so heavily dependent of the subject as a self-transparent being.*

This postmodern understanding of the subject, as a decentered self, initiated by scholars such as Freud, may be viewed as lying at the heart of much of the contemporary constructionist critique of traditional social psychology and consumer behaviour research. Firat and Venkatesh (1995: 243), among others, have argued that the human subject (or self) may better be understood as fragmented because in the micropractices of everyday life, discontinuities, pluralities, chaos, instabilities, constant changes, fluidities, and paradoxes define the human condition. Therefore, in such a fragmented society where consumers face a bewildering range and fluidity of social roles, values and authorities, traditional theoretical models that tend to reduce consumer behaviour to observable signs of fixed and stable psychological tendencies are losing their appeal. As Hall (1996: 226) has argued,

> *We can no longer conceive of the 'individual' in terms of a whole, centred, stable and completed Ego or autonomous, rational 'self'. The 'self' is conceptualized as more fragmented and incomplete, composed of multiple 'selves' or identities in relation to the different social worlds we inhabit, something with a history, 'produced', in process. The 'subject' is differently placed or positioned by different discourses and practices.*

From this perspective, the widely adopted assumption – and demand – in the social marketing literature on green consumerism that a more or less conscious rational processing lies behind or guides green consumers' outward behaviour should be significantly relaxed. More attention should be paid to how consumers "do rationality" (Gergen and Joseph, 1997). For a green consumer, representing oneself as a rational moral actor arguably involves participation in a discursive system that is already constituted; it involves borrowing from the existing idioms, appropriating forms of talk and related action already in place. Thus, as Gergen and Joseph put the point, rationality is not to exercise an obscure and interior function of 'thought' but to participate in a form of cultural life or to play by the rules favored within a particular tradition. In the myriad social contexts that everyday life provides, these rules of rationality may well be very different. Moral justifications that seem 'rational' when filling in survey questionnaires focusing on 'environmental concern' may not be so 'rational' in other social situations and discursive contexts. Therefore, grounding research in theories that insist more or less explicitly on some kind of rationality of moral reasoning does not seem a very useful strategy for investigating the various moral dimensions of green consumerism.

Unsurprisingly, then, Harré et al. (1999), in their study on 'Greenspeak'[6], found a multitude of different and often divergent positions, currents and views. Therefore, a truly descriptive approach to studying the moral dimensions of green consumerism, focusing on the origin and construction of the meaning of moral norms, ideals and principles for individual consumers, as well as on the discursive aspects of green consumerism as a social movement, would seem to offer valuable insights into the social and behavioural dynamics of ecologically responsible behaviour. As Harré et al (1999: 183) put the point, "the use of moral concept is to be investigated, not taken for granted".

5.3 PERSONAL ENVIRONMENTAL ETHICS AS AN EXISTENTIAL CHOICE

Instead of postulating a 'green consumer ethics' based on rational, teleological reasoning or on simple environmental codes of conduct, personal environmental ethics may be viewed as an existential choice (Harré *et al.*, 1999) or as an 'aesthetic of existence' (Darrier, 1999). Harré et al. (1999), have observed that for many people being an environmentalist is primarily an existential choice. They point out that

> *to be a Greenspeaker is not only to have adopted wise environmental practices as measures of self-defense or for some wider utilitarian motive or to have invested one's choices with moral value but to a be a certain kind of person* (ibid. p. 180).

In much the same vein, Darier (1999), drawing on Foucault (1984), suggests an approach to environmental ethics that he calls an 'aesthetic of existence', or 'arts of existence'. Darier (1999:238) argues that

---

6 Harré et al. (1999: vii) use the term 'greenspeak' to refer to the ways in which issues of the environment are presented in text, talk, and images.

> *...reducing individual energy conception in the North should not be justified by an imperative / threat like 'global warming' defined by an 'expertocracy'..., but because one might not want the consumption of large amounts of energy to be a defining characteristic of oneself!*

Darrier (1999), among others, is suspicious of most conventional 'environmental ethics' because their advocates tend to use naturalistic and moralistic justifications, thus falling into the naturalistic fallacy[7]. Recurrently, the legitimation of most contemporary environmental ethical theories seems to rest on a belief that natural science and the 'natural world' should be the source of norms and moral principles for people to obey (Darrier, 1999; Hannigan, 1995; Harré *et al.*, 1999; Meriläinen *et al.*, 2000). A study on 'Greenspeak' by Harré et al. (1999), for example, shows that natural sciences constitute a "rich reservoir of terminology and models" and a powerful source of rhetorical devices that environmentalist use in shaping public opinion (ibid. p. 67). Hence, as Darier (1999: 217) has put the point, "[l]ike gods and 'objective scientific truth', 'nature' becomes another normative yardstick to impose itself on human behaviour and values." In the name of the presumed proper functioning of the ecosystem, for example, people are advised to adopt new norms, values, behaviours and solutions, which in much the same way as the laws of nature apply universally, transcending the cultural and the historical.

It may be argued, however, that since the validity or 'truth' of scientific knowledge is context specific, emerging from specific worldviews and discourses or from specific paradigmatic communities (Harding, 1993, 1995; Longino, 1990), there is no reason to expect that natural sciences could offer clear, certain, absolute, and universal sets of environmental ethical norms (Darier, 1999). Since all environmental knowledge is inevitably intertwined with power relations, it is impossible to articulate an environmental ethic that would be free from existing relations of power. (ibid. p. 219, also Braidotti *et al.*, 1994) Indeed, a number of scholars have observed that traditional and contemporary environmentalism, in the many forms it takes, shares many features with the power they wish to oppose (Darier, 1999; Harré *et al.*, 1999, Meriläinen *et al.*, 2000). Consequently, environmental ethics cannot adequately be conceived in terms of grand universal principles detached from the context of their actual production, as Darier (1999: 220) points out.

To fend off the "naturalistic fallacy", Darier (1999: 226) proposes an "environmental ethic à la Foucault" based on a "green aesthetic of existence". Such an environmental ethics is grounded in constant self-reflection, self-knowledge, and self-examination of the existing limits of what constitutes the environment and the individual's conduct toward the environment and toward oneself. Instead of offering simple solutions to the environmental crisis, it merely suggest the adoption of 'ethical sensibility'. Such sensibility is grounded upon a constant critical, skeptical attitude toward foundationalism and assumed natural categories. It involves a continued

7 'Naturalistic fallacy', in Darier's (1999: 217) argumentation, refers to the use of naturalistic and moralistic justifications for legitimating environmental ethical theories. He questions the idea that the alleged 'natural world' should be the source of norms or directions that humans should obey.

transgression of the limits of the conditions that have constructed our current and past subjectivities, i.e., a permanent reinventing of ourselves.

As a result, according to Darier, the green aesthetic of existence leads to reflective and critical 'environmental resistance': to radical questioning of the existing dominant discourse and practices around 'nature', as well as to practices of transgression of the given subjectivity of consumer. Through this process of constant hyper-criticism one can question the conditions which account for one's subjectivity, and start to imagine and build new kinds of subjectivities.

Moreover, the green aesthetic of existence acknowledges the impossibility of grounding environmental ethics upon absolute, external, universal and coherent standards or a 'truth' about 'nature'. According to Darier (1999: 233-234) it is not possible to evaluate in the abstract the degree of 'greenness' of any behaviour because there are no absolute external referential categories for greenness. Since all actions are situated in a specific context of power relations, and because of the extreme fluidity and adaptivity of the relations of power, a genuine act of green resistance, at a given point of time, can over time become one that merely legitimizes the existing system of power relations.

Consequently, green ethics as an aesthetic of human existence is rooted in a permanent radical questioning and re-questioning of the broader conditions that result in consumers seeing the world as they see it. This sort of personal environmental ethics encourages consumers to promote new forms of subjectivity through the refusal of the kind of individuality that has been imposed on citizens as 'consumers' in the history of Western market economies. Instead of asking what can we know and how should we behave, it might be more fruitful to ask "how have my questions been produced?" and "How have I been situated to experience the real?" (Bernauer, 1992, ref. Darier 1999: 224).

5.4. SUGGESTED FUTURE DIRECTIONS FOR ENVIRONMENTAL POLICY ORIENTED RESEARCH

5.4.1. A constructionist alternative to personal ethics and green consumerism

This paper has aimed to argue that, in social marketing, it would be useful to investigate green consumerism as a social and cultural phenomenon. For a deeper and more thorough understanding of social and behavioural dynamics of environmentally responsible consumer behaviour, it would seem important to study the social construction of the identity positions and moral norms associated with 'green consumerism'. This would seem to require a truly descriptive approach, focusing on culturally shared ideas, meanings and signifying practices that constitute the dominant discourses on green consumerism.

The starting point for such constructionist research is the idea that consumers' preferences and values are not stable and do not arise solely from biological needs, essential psychological characteristics or universal mental processes, but are socially constructed in their daily interaction processes. It is thus emphasized that producers, marketers, and consumers produce moral ideals, needs and wants jointly in interaction. Therefore, rather than trying to 'measure' the content and structure of consumers' personal environment-related moral beliefs, ideas and values, it would seem necessary and interesting to focus research on the processes in which these beliefs, ideas and values are produced, strengthened and changed. Such an approach to social marketing questions the individualist conception of the subject, and the ideals of autonomy, rationality, and separateness, and emphasizes the social embeddedness and contingency of human behaviour. The point of view thus shifts from persuasion and production of images, in which unreflectively pre-defined representations and moral norms are imposed on consumers, to critical questioning of the dominant discourse on the environment and to the creation and co-production of new, more sustainable culturally shared meanings and subject positions for the 'consumer'.

A strong identification with 'green consumerism' is undoubtedly important for assuming personal responsibility for environmental protection because such an identity position helps consumers to find environmentally responsible behaviours meaningful. However, it is emphasized that social life in contemporary Western market economies, with its multiplicity of social contexts in everyday life, offers consumers multiple identity possibilities, the identity of 'green consumer' being only one of them. People do not live in social isolation – although some mainstream models of consumer behaviour implicitly make this idealization. Therefore, 'identity' can be viewed as a project, developed from multiple sources and collective understandings in local and global communities (Hall, 1992; Stevens and Wetherell, 1996; Wetherell, 1996). Nevertheless, this project may be based on more or less unconscious attempts to try to develop unified narratives about all the diverse relationships and activities they engage in. A narrative of a green consumer may well be one of them. It is necessary to note, however, that this attempt to unify or make sense of one's own identity involves struggle, repression, and internal conflict. There are no grounds for assuming that it is based on some whole, centered, stable, and completed ego or independent and self-gratifying rational 'self', as discussed earlier. Consequently, when studying green consumerism from a constructionist perspective, it would seem necessary to shift the focus away from "the presumed discovery or 'rediscovery' of a true permanent ecological self to the active constitution of subjectivities which constantly rework humans' relations with themselves, with other life-forms, and with the world generally", as Darier (1999: 27) has put the point.

5.4.2. A potential contribution of constructionist social marketing to sustainable development

Overall, this paper has aimed to discuss the limitations of the dominant approach to environmental policy oriented research on green consumerism, arguing that it is flawed, to some extent, because it is largely based on what has been called 'individualist' and 'experimentalist' assumptions of subjectivity and human behaviour. Framing green consumerism in terms of individual choice and morally responsible decision-making, this approach has tended to produce solutions for tackling the human dimensions of environmental problems that are predominantly techno-economic, and largely also neo-classically informed (Dennis *et al.* 1990; Kempton et al., 1990; Mc Mahon, 1997; Mellor, 1997; Stern, 1992). Therefore, it is argued here that to realize the potentially significant contribution that social marketing research may offer to suctainable development, it seems necessary that the study of green consumerism be framed in terms of conceptual frameworks that are essentially different from those informed by experimentalist social psychology and neo-classical microeconomics.

Following Gergen and Joseph (1997), it may be argued that the concepts, descriptions, explanations, technologies and images that environmental policy oriented social marketing research produces may be viewed as a significant generative source of meanings in cultural life. In generating and disseminating these 'meanings', it provides people with pragmatic devices through which organizational, political and cultural life is carried out. As Gergen (1997) has pointed out, professional accounts, bearing the stamp of scientific authority, inform people's actions and instruct social policy. In doing this, social marketing inherently operates to the benefit of certain stakeholders, activities, and forms of cultural life and to the detriment of others. Therefore, one of the main challenges of constructionist social marketing would be to engage in a reflexive deliberation – perhaps ideological unmasking – of the taken-for-granted views on environmental problems and human behaviour. This would involve not only an examination of the cultural implications of the constructions of social marketing, but also active participation in the more general debates about values and goals within the culture, especially as these are related to organizational and business practices (Gergen and Joseph, 1997). In particular, constructionist social marketing research would seem to have a promising potential contribution in critically examining and ideologically unmasking the implicit assumptions and their political implications of the dominant neo-classically informed, market-based environmental policy programs.

However, as Gergen (1997) has argued, the critical voice of the constructionists should not be viewed as liquidating but as serving the useful functions of denaturalization and democratization. He argues that by denaturalizing the 'objects of research', methodologies, and resulting practices, such critical inquiry invites an appropriate humility and fosters and favors pluralist politics, both within the profession and with respect to the profession's relationship to its many publics, thus opening the profession to multiple voices from the culture more generally.

Moreover, such reflexive critique within the profession also invites the social marketing scholars to enter the process of creating realities by skillfully fashioning its images, concepts, metaphors, narratives, and discourses. Informed by Gebgen and Joseph (1997), it may be observed that whereas the traditional social marketing research on green consumerism has largely focused on charting existing patterns of behaviour, a constructionist approach to environmentally responsible consumption is challenged to break the barriers of common sense by offering new forms of theory, of interpretation, or intelligibility. Since the task of an experimentalist researcher is primarily to give accurate accounts of existing reality, taken to be an instantiation of universal and transhistorical processes, the experimentalist scholars have taken little interest in molding new futures for society, as Gergen (1996) has pointed out. Moreover, the received ideals of 'objectivity' and 'value neutrality', may have inhibited experimentalist social marketing scientists from pursuing social change. Consequently, there appears to be a need for more "generative" theorizing (Gergen, 1994) on the social and behavioural dynamics of green consumerism: theorizing designed to unseat conventional assumptions, and to open new alternatives for action (Gergen and Joseph, 1997).

In conclusion, I would argue that there is a need for constructionist accounts of the social and behavioural dynamics of environmental problems that question, perhaps radically, the dominant techno-economic discourse on sustainable development and climate mitigation, and open new alternative views and perspectives to the human dimensions of environmental policy. A valuable potential contribution that social marketing research, and social science in general, can make to sustainable development and environmental policy would seem to be to feed into and to bring about what Darier (1999) calls 'critical environmental resistance', or what Gergen and Joseph (1997) call 'reflexive deliberation'. A major task and challenge for social marketing research, then, would be to engage in radical questioning of the existing dominant techno-economic discourse and practices around 'nature', exploring, building and introducing new kinds of subject positions for citizens as consumers.

6. Appendix

Examples of mainstream marketing studies on environmentally responsible behaviour: articles published in the Journal of Macromarketing and in the Journal of Public Policy and Marketing in the 1990's.

Author/ Year	**Framing and problematics**	**Conceptualization of environmental concern/ green consumerism**	**Motivation**
Fitz-gerald & Corey (1992)	Ethical choice as a morally reasoned justification for choice of action Revealing "ethics gaps", ethical dilemmas Survey	judgment as ethical or unethical as regards using non-degradable packaging and packaging hazardous products in recyclable materials	responsibility towards different stakeholders deontological ethics
Pelton et al. (1993)	Ethical decision-making as altruistic behaviour Investigating the decision criteria that govern consumers' willingness to recycle Survey	willingness to recycle, attitudes as teleological evaluations toward recycling	utility maximizing behaviour weighting egoistic and utilitarian decision outcomes teleological ethics
Carman (1992)	Prosocial behaviour/ altruistic behaviour as regard for others, co-operation for the good of self and others Conceptual analysis of past literature explaining the decline in individual	energy conservation, increasing energy efficiency, curtailing use	utility maximizing: private utility vs. public utility teleological ethics

	altruism		
Wiener (1993)	Cooperation in a social dilemma Investigating the willingness of a natural decision making group to impose a collective sacrifice on themselves to solve a social dilemma, associated with energy conservation Experiment with a student sample	willingness to agree on the installation of load control devices on the air conditioners cooling university class rooms	utility maximizing behaviour collective vs. individual rationality teleological ethics
Granzin & Olsen (1991)	Helping behaviour to find variables useful for characterizing citizens who differ in terms of their willingness to participate in environmental activities survey	participation in 3 environmental protection activities donating items for reuse, recycling newspapers, and walking when possible for reasons of conservation and environmental concern	internalized responsibility, personal obligation to help owing to intrinsic satisfaction from helping and approval from others valuing helpfulness and altruistic values teleological ethics
Pieters (1991)	goal attainment and rational decision-making literature review analyzing factors influencing participation in waste separation programs	continuing participation of consumers in 19 waste separation programs	subjective expected utility of behavioural performance or goal attainment: perceived personal costs and benefits teleological ethics
Schwep-ker & Cornwell (1991)	Social responsibility, (helping others even when there is nothing to be gained) Predicting ecological concern :isolating useful variables for identifying the ecologically concerned consumer Survey	willingness to purchase ecologically packaged products	adhering to social norms and accepted social values deontological ethics
Singhapakdi & LaTour (1991)	Socially responsible consumption orientation the effects of altruism, utilitarianism and social responsibility on voting intentions experiment	voting intention to support a container deposit law	social responsibility, concern for the welfare of society in general environmental vs. personal benefits teleological ethics

Examples of mainstream marketing studies on environmentally responsible behaviour: articles published in the Journal of Macromarketing and in the Journal of Public Policy and Marketing in the 1990's. (cont.)

Author/ Year	**Framing and problematics**	**Conceptualization of environmental concern/ green consumerism**	**Motivation**
Powers et al. (1992)	Informed consumption choices as an actor in market exchange system under constraints imposed by the social structure the influence of demographics on participation in energy conservation programs survey	taking an energy audit by electric utility and making improvements	energy conservation
Taylor & Todd (1995)	Attitude-behaviour relationship the relative importance of attitudes, norms and PBC on individual waste management behaviour survey + diary	household garbage reduction	evaluations of the outcomes of behaviour personal vs. societal advantages
Pilling et al. (1991)	Attitude-behaviour relationship Application of benefit segmentation to a public policy issue Survey	voting behaviour concerning a container law	evaluations of the outcomes of behaviour
Ellen et al. (1991)	Attitude-behaviour relationship can perceived consumer effectiveness be empirically distinguished from environmental concern the effects of PCE on individual efforts and joint efforts of many individuals telephone survey	environmentally conscious action as an index of activities reflecting different degrees of individual effort in dealing with environmental problems:	awareness of environmental problems and a high priority on solving environmental problems
Pickett et al. (1993)	Autonomous choice of behaviour based on rational evaluation of the desirability of the behaviour the influence of demographic and psychosocial factors on ecologically beneficial behaviour	a pattern of ecologically beneficial behaviour: garbage separation and recycling, using reusable containers to store food, conserving water when doing dishes, switching of lights, not throwing away durables	social responsibility, pro-social behaviour (not defined)

Berger & Kanetkar (1995)	Attitude-behaviour relation, personal consumption choices Whether and when consumers will sacrifice personal benefits for societal considerations Naturally occurring quasi experiment	expressing high levels of sensitivity to environmental attributes in choosing a detergent	concern about the state of the physical environment evaluation of the current state of his or her physical habitat
Pieters et al. (1998)	Choices in social dilemma situations Examine attributions that consumers make about the proenvironmental behaviour, motivation and ability of relevant societal actors Computerized survey	an index of proenvironmental behaviours regarding energy and water use, recycling, purchasing, and disposing	awareness of environmental problems and willingness to place high value on solving these problems
Hutton & Ahtola (1991)	Attitude-behaviour relation, adoption and diffusion of a social innovation Assess the impacts of a 5-year, naturally evolving social marketing program to reduce carbon monoxide Quasi-experiment, survey	participation in the program; willingness to sacrifice the current benefits of driving not driving, reducing driving, not burning wood, using oxygenated fuels	social traps: utility maximizing behaviour individual responsibility for clean air vs. subjective costs or sacrifice
Apaiwongse (1991)	Adoption of an industrial innovation Survey	industrial buying center members' attitudes toward adopting an environmentally related public policy	perceived benefits of the policy vs. perceived uncertainty in decision-making

7. References

Ajzen, I. and Fishbein M., 1980: *Understanding attitudes and predicting social behaviour.* Prentice-Hall, Englewood Cliffs.

Andreasen, A. R., 1991: Consumer Behaviour Research and Social Policy. In *Handbook of Consumer Behaviour.* [Robertson, T. S. and H. H Kassarjian (eds.)]. Prentice-Hall, Englewood Cliffs, pp.459-506.

Andreasen, A. R., 1994: Social Marketing: Its Definition and Domain. *Journal of Public Policy and Marketing*, 13(1), pp. 108-114.

Anderson, W. T. and W. H. Cunningham, 1972: The Socially Conscious Consumer. *Journal of Marketing*, 36, pp. 23-31.

Antil, J. H., 1984: Socially Responsible Consumers: Profile and Implications for Public Policy. *Journal of Macromarketing*, 4(2), pp. 18-39.

Antil, J. H. and P. D. Bennet, 1979: Construction and Validation of a Scale to Measure Socially Responsible Consumption Behaviour. In *The Conserver Society.* [Henion, K. E. and T.C. Kinnear (eds.)]. American Marketing Association, Chicago, pp. 51-68 .

Antonides, G. and W. F. van Raaij, 1998: *Consumer Behaviour: A European Perspective*. John Wiley & Sons, Chichester:

Banerjee, S. and C. S. Gulas, and E. Iyer, 1995: Shades of Green: A Multidimensional Analysis of Environmental Advertising. *Journal of Advertising,* 24(2), pp. 21-31.

Belk, R. W., 1987: A modest Proposal for Creating Verisimilitude in Consumer Information Processing Models and Some Suggestions for Establishing a Discipline to study consumer Behaviour. In: *Philosophical and Radical Thought in Marketing* [Firat, Fuat A. and Nikhlesh Dholakia, and Richard Bagozzi (eds.)].. Lexington Books, Lexington, pp. 361-384.

Bell, A., 1994: Climate of opinion: public and media discourse on the global environment. *Discourse & Society,* 5(1), pp. 33-64.

Berger, I.E. and V. Kanetkar, 1995: Increasing Environmental Sensitivity Via Work Place Experiences. Journal of Public Policy and Marketing, 14 (2), pp. 205-215.

Bernauer, J.W., 1992: Beyond Life and Death: On Foucault's Post-Auschwitz Ethic. In: *Michel Foucault Philosopher* [Armstrong (trans.)]. Harvester Wheatsheaf, Hemel Hempstead, pp. 260-279.

Braidotti, R., E. Charkiewicz, S. Häusler, and S. Wieringa, 1994: *Women, the Environment and Sustainable Development: Towards a Theoretical Synthesis.* Zed Books, London.

The Cambridge Dictionary of Philosophy, 2nd edition [Audi, R. (ed.)], Cambridge University Press, Cambridge 1999.

Carman, J.M., 1992: Theories of Altruism and Behaviour Modification Campaigns. *Journal of Macromarketing*, 12(1), pp. 5-18.

Chaiken, S. and C. Stangor, 1987: Attitudes and Attitude Change. In: *Annual Review of Psychology*, Vol. 38 [Rosenzweig, Mark R. and Lyman W. Porter (eds.)]. Annual Reviews, Palo Alto CA, pp. 575-630.

Darier, E., 1999: Foucault against environmental ethics. In: *Discourses of the Environment.* [Darier, E. (ed.)]. Blackwell Publishers, Oxford, pp.217-240.

Dawes, R., 1980: Social dilemmas. *Annual Review of Psychology*, 31, pp. 69-93.

Dawes, R.M., A.J.C. van den Krag, and J.M. Orbell, 1988: Not Me or Thee but We: The Importance of Groep Identity in Dilemma Situations: Experimental Manipulations. *Acta Psychologica*, 68, pp. 83-97.

Dennis, M.L, E.J. Soderstrom, W.S. Koncinski, Jr., and B. Cavanaugh, 1990: Effective Dissemination of Energy-Related Information: Applying Social Psychology and Evaluation Research. *American Psychologist*, 45(10), pp. 1109-1117.

Elkington, J., J. Hailes, and J. Makower, 1990: *The Green Consumer.* Penguin Books, New York.

Ellen, P.S., J.L. Wiener, and C. Cobb-Walgren, 1991: The Role of Perceived Consumer Effectiveness in Motivating Environmentally Conscious Behaviours. *Journal of Public Policy and Marketing*, 12(2), 102-117.

Firat, F. A., N. Dholakia, and R. Bagozzi, 1987: Introduction: Breaking the Mold. In: *Philosophical and Radical Thought in Marketing* [F. A. and N. Dholakia, and R. Bagozzi (eds.)]. Lexington Books, Lexington. pp. xiii-xxi.

Fishbein, M. and Ajzen, I., 1975: *Belief, attitude, intention and behaviour: an introduction to theory and research.* Addison-Wesley, Reading.

Fjeld, C. R., H.G. Schultz, and R. Sommer, 1984: Environmental Values of Food Cooperative Shoppers. In: *Personal Values and Consumer Psychology* [Pitts, R. and Woodside, A. (eds.)]. Heath and Co, Massachusetts: D.C, pp. 287-295.

Foucault, M., 1984: *The Use of Pleasure*. Pantheon Books, New York.

Frank, T.H., 1988: *Passions within Reason: The Strategic Role of the Emotions*. W.W. Norton, New York.

Fuller, D.A., 1999: *Sustainable Marketing*, Sage Publications, London.

Fuller, D.A., J. Allen, and M. Glaser, 1996: Materials Recycling and Reverse Channel Networks: The Public Policy Challenge. *Journal of Macromarketing*, 16(1), pp. 52-72.

Gardner, G.T. and P.C. Stern, 1996: *Environmental Problems and Human Behaviour*. Allyn and Bacon, Boston.

Ger, G., 1999: Experiential Meanings of Consumption and Sustainability in Turkey. *Advances in Consumer Research*, 26, pp. 276-280.

Gergen, K.J., 1996: Social Psychology as Social Construction: The Emerging Vision. In: *The Message of Social Psychology: Perspectives on Mind in Society* [McGarty, C. and S. A. Haslam (eds.)]. Blackwell Publishers, Oxford, pp. 113-128.

Gergen, K.J., and T. Joseph, 1997: Organizational Science in a Postmodern Context. *Journal of Behavioural Science*, 32, pp. 356-377.

Gergen, K.J., 1997: The Place of the Psyche in a Constructed World. *Theory and Psychology*, 7, 723-746.

Gilroy, P., 1997: Diaspora and the Detours of Identity. In: *Identity and Difference [*Woodward, K. (ed.)]. Sage Publications/The Open University, London , pp. 299-343.

Granzin, K.L. and J.E. Olsen, 1991: Characterizing Participants in Activities Protecting the Environment: A Focus on Donating, Recycling, and Conservation Behaviours. *Journal of Public Policy and Marketing,* 10(2), pp. 1-27.

Grosz, E A., 1990: *Jacques Lacan - A Feminist Introduction.* Routledge, London.

Hackley, C.E., 1998: Social constructionism and research in marketing and advertising. *Qualitative Market Research: An International Journal*, 1(3), pp. 125-131.

Hackley; C. and P. Kitchen, 1987: IMC: a consumer psychological perspective. *Marketing Intelligence & Planning*, 16(3), pp. 229-235.

Hall, S., 1997: The Work of Representation. In: Hall, S. (ed.) *Representation: cultural representations and signifying processes,* pp. 13-74.

Hall, S., 1996: The meaning of new times. In: *Stuart Hall: critical dialogues in cultural studies* [D. Morley and K-H Chen,]. Routledge, London, pp. 223-237.

Hall, S., 1992: The Question of Cultural Identity. In: *Modernity and Its Futures* [Hall, S. and D. Held and T. GcGrew (eds.)]. Sage Publications/ The Open University, London, pp. 273-316.

Hannigan, John A., 1995: *Environmental Sociology - a social constructionist perspective*, London: Routledge.

Hardin, G., 1968: The Tragedy of the Commons. *Science*, 162, pp. 1243-1248.

Harding, Sandra, 1993: Rethinking Standpoint epistemology: What is 'Strong Objectivity'? In: *Feminist Epistemologies* [Linda Alcoff and Elisabeth Potter (eds.)]. Routledge, New York, pp. 49-82.

Harding, S., 1995: Can Feminist Thought Make Economics More Objective? *Feminist Economics*, 1, pp. 7-32.

Harré, R., J. Brockmeier, and P. Mülhäuser, 1999: *Greenspeak: A Study of Environmental Discourse,* Sage Publications, Thousand Oaks, CA.

Heiskanen, E. and Pantzar, M., 1997: Toward Sustainable Consumption: Two New Perspectives. *Journal of Consumer Policy*, 20, pp. 409-442.

Henion, K., E., 1976: *Ecological Marketing,* Columbus, Grid Inc Ohio.

Hollis, M., 1994: *The Philosophy of Social Science.*: Cambridge University Press, Cambridge.

Hutton, R. B. and O.T. Ahtola, 1991: Consumer Response to a Five-Year Campaign To Combat Air Pollution. *Journal of Public Policy and Marketing*, 10(1), pp. 242-256.

Jencks, C., 1990: Varieties of Altruism. In: *Beyond Self Interest* [Mansbridge, J. (ed.)]. The University of Chicago Press, Chicago, pp. 54-67.

Joy, A. and A. Venkatesh, 1994: Postmodernism, feminism, and the body: The visible and the invisible in consumer research. *International Journal of Research in Marketing*, 11, pp. 333-357.

Kalafatis, S.P., M. Pollard, R. East, and M.H. Tsogas, 1999: Green marketing and Ajzen's theory of planned behaviour: a cross-market examination. *Journal of Consumer Marketing*, 16(5), pp. 441-460.

Kempton, W., 1991: Lay Perspectives on Global Climate Change. In: *Energy Efficiency and the Environment: Forging the Link [*Vine, E. and D. Crawley and P. Centolella (eds.)]. American Council for Energy Efficient Economy, Berkeley.

Kempton, W., J.M Darley, and P.C. Stern, 1992: Psychological Research for the New Energy Problems. *American Psychologist*, 47(10), pp. 1213-1223.

Kumcu, E., 1987: Historical Method: Toward a Relevant Analysis of Marketing Systems. In: *Philosophical and Radical Thought in Marketing* [F. A. and N. Dholakia, and R. Bagozzi (eds.)]. Lexington Books, Lexington. pp.117-133.

Kotler, P. and E. Roberto, 1989: *Social Marketing: Strategies for Changing Public Behaviour.* The Free Press, New York.

Lalljee, M., 1996: The Interpreting self: An experimentalist perspective. In: Stevens, Richard (ed.) *Understanding The Self.* Sage Publications, London, pp. 89-145.

Longino, H. E., 1990: *Science as Social Knowledge: Values and Objectivity in Scientific Inquiry.* Princeton University Press, Princeton.

Mappes, T.A. and J.S. Zembaty, 1992: *Social Ethics: Morality and Social Policy*, Fourth Edition, McGraw-Hill, New York.

McMahon M., 1997: From the ground up: ecofeminism and ecological economics. *Ecological Economics*, 20, pp. 163-173.

Mellor, M., 1997: Women, nature and the social construction of 'economic man'. *Ecological Economics*, 20, pp. 129-140.

Meriläinen, S., J. Moisander, and S. Pesonen, 2000: The Masculine Mindset of Environmental Management and Marketing. *Business Strategy and the Environment*, 9(3), pp. 151-162.

Milbrath, L.W., 1986: Environmental Beliefs and Values. In: *Political Psychology* [Herman, M. (ed.)]. Jossey-Bass Publishers, San Francisco, pp. 97-138.

Moisander, J., 1991: *Sosiaaliset arvot luomutuotteiden kulutuksessa* (In Finnish, Social values and the consumption of organic food products), Publications of Helsinki School of Economics and Business Administration D-144, Helsinki.

Moisander, J., 1996: *Attitudes and Ecologically Responsible Consumption. Moral responsibility and concern as attitudinal incentives for ecologically oriented consumer behaviour*. Research Reports 218/1996. Statistics Finland, Helsinki.

Moisander, J., 1997: Gender Dynamics in Green Consumption: Analyzing and Exploring the Interconnections Between Green Consumerism and Female Subjectivity." In: *Proceedings of the XXII International Colloquium of Economic Psychology*, Vol. I, pp. 357-366.

Naess, Arne, 1989: *Ecology, Community and Lifestyle*. Cambridge University Press, New York.

Nelson, J. A., 1997: Feminism, ecology and the philosophy of economics, *Ecological Economics*, 20, pp. 155-162.

Osterhus, T., 1997: Pro-Social Consumer Influence Strategies: When And How Do They Work? *Journal of Marketing*, 61, pp. 16-29.

Pelton, Lou E., David Strutton, James H. Barnes Jr., and Sheb L. True, 1993: The relationship among Referents, Opportunity, Rewards, and Punishments In: Consumer Attitudes toward Recycling: A Structural Equations Approach. *Journal of Macromarketing*, 13(1), pp. 60-74.

Peattie, K., 1992: *Green Marketing*. Pitman Publishing, London.

Peñaloza, Lisa, 1994: Crossing boundaries/drawing lines: A look at the nature of gender boundaries and their impact on marketing research. *International Journal of Research in Marketing*, 11, 359-279.

Pieters, R.G.M., 1991: Changing Garbage Disposal Patterns of Consumers: Motivation, Ability and Performance. *Journal of Public Policy and Marketing,* 10, pp. 59-76.

Pieters, R., T. Bijmolt, F. van Raaij, and M. de Kruijk, 1998: Consumers' Attributions of Proenvironmental Behaviour, Motivation, and Ability to Self and Others. *Journal of Public Policy and Marketing*, 17(2), pp. 215-225.

Pilling, B. K., L. A. Crosby, and P. S. Ellen, 1991: Using Segmentation to Influence Environmental Legislation: A Bottle Bill Application. *Journal of Public Policy and Marketing*, 10(2), pp. 28-46.

Potter, J., 1996: Attitudes, social representations and discursive psychology. In: *Identities, Groups and Social Issues* [Wetherell, M. (ed.)]. Open University/Sage Publications, London , pp.119-173.

Potter, J. and M. Wetherell, 1987: *Discourse and Social Psychology*. Sage Publications, London.

Powers, T.L., J.E. Swan and S.-D. Lee, 1992: Identifying and Understanding the Energy Conservation Consumer: A Macromarketing Systems Approach. *Journal of Macromarketing*, 12(2), pp. 5-15.

Prior, L., 1997: Following Foucault's Footsteps: Text and Context in Qualitative Research. In: *Qualitative Research: Theory, Method and Practice* [Silverman, D. (ed.)]. Sage Publications, London, pp. 63-79.

Puranen, Bi, 2000: The challenge of climate change trough the eyes of the Swedish youth. In: *Climate Change: socioeconomic dimensions and consequences of mitigation measures* [Pirilä, Pekka (ed.)]. Fortum and Edita, Helsinki, pp. 335-354.

Roper Starch Worldwide Inc., 1996: *Green Gauge Report.*

Singhapakdi, A. and M.S. LaTour, 1991: The Link Between Social Responsibility Orientation, Motive Appeals, and Voting Intention: A Case of an Anti-littering Campaign. *Journal of Public Policy and Marketing*, 10(2), pp. 118-129.

Schwepker, C.H. Jr. and B. Cornwell, 1991: An Examination of Ecologically Concerned Consumers and Their Intention to Purchase Ecologically Packaged Products. *Journal of Public Policy and Marketing*, 10(2), pp. 77-101.

Shrum, L. J., J.A. McCarty, and T.M. Lowrey, 1995: Buyer Characteristics of the Green Consumer and Their Implications for Advertising Strategy. *Journal of Advertising*, 24(2), pp. 71-82.

Sparks, P. and S. Richard, 1992: Self-Identity and the Theory of Planned Behaviour: Assessing the Role of Identification with Green Consumerism. *Social Psychology Quarterly,* 55, pp. 388-399.

Steer, A., 1996: Ten principles of New Environmentalism. *Finance & Development*, 33(4), pp. 4-7.

Stern, P.C., 1992: What Psychology Knows about Energy Conservation. *American Psychologist*, 47(10), pp. 1224-1232.

Stevens, R., 1996: Introduction: Making sense of the person in a social world. In: *Understanding The Self* [Stevens, R. (ed.)]. Sage Publications, London., pp. 1-34.

Stevens, R. and M. Wetherell, 1996: The self in the modern world: drawing together the threads. In: *Understanding the Self* [Stevens, R. (ed.)]. Sage Publications/The Open University, London , pp. 339-369.

Taylor, S. and P. Todd, 1995: Understanding Household Garbage Reduction Behaviour: A Test of an Integrated Model. *Journal of Public Policy and Marketing*, 14(2), pp. 192-204.

Thompson, C.J. 1993: Modern truth and postmodern incredulity: A hermeneutic deconstruction of the metanarrative of "scientific truth" in marketing research. *International Journal of Research in Marketing*, 10, pp. 325-338.

Thørgesen, J., 1994: A model of recycling behaviour, with evidence from Danish source separation programmes. *International Journal of Research in Marketing*, 11, pp. 145-163.

Turner, R. K., D. Pearce and I. Bateman, 1994: *Environmental Economics*. Harvester Wheatsheaf, Hertfordshire.

Uusitalo, L., 1989: Economic Man or Social Man - Exploring Free Riding in the Production of Collective Goods. In: *Understanding Economic Behaviour* [Grunert, K.G and F. Ölander (eds.)]. Kluwer Academic Publishers, Dordrecht, pp. 67-283.

Uusitalo, L., 1991: Dilemma between individual utility seeking and collective welfare seeking behaviour. *Archiv für Rechts- und Sozialphilosophie*, Beiheft 40, [Eugene E. Dais, Stig Jørgensen, Alice Erh-Soon Tay (Hrsg.)]. Franz Steiner Verlag, Stuttgart, pp. 181-191.

Uusitalo, L. and J. Uusitalo, 1980: Scientific Progress and Research Traditions in Consumer Research, With Special Reference to Consumption Related Problems. *Advances in Consumer Research*, 8.

Vlosky, R.P., L.K. Ozanne, and R. Fontenot, 1999: A conceptual model of US consumer willingness-to-pay for environmentally certified wood products. *Journal of Consumer Marketing*, 16(2), pp. 122-136.

Webster, F.E., 1975: Determining the Characteristics of the Socially Conscious Consumer. *Journal of Consumer Research*, 2, pp. 188-196.

Wetherell, M., 1996: Group conflict and the social psychology of racism. In: *Identities, Groups and Social Issues* [Wetherell, M. (ed.)]. Open University/Sage Publications, London, pp. 239-298.

Wiener, J.L, 1993: What Makes People Sacrifice Their Freedom for the Good of Their Community? Journal of Public Policy and Marketing, 12(2), pp. 244-251.

Wiener, J.L. and T.A. Doescher, 1991: A Framework for Promoting Co-operation. *Journal of Marketing*, 55, pp. 38-47.

Zinkhan, G.M. and L. Carlson, 1995: Green Advertising and the Reluctant Consumer. *Journal of Advertising* , 24(2), pp. 1-6.

European Narratives about Human Nature, Society and the Good Life

Laurie Michaelis

A13
I31 Z13

1. Introduction

Other papers in this book (Ney/Thompson 2000; Thompson, 2000) discuss the role of narrative and discourse in shaping our interpretation of, and response to, climate change. Analysis of discourses is an important tool in making sense of political issues and debates. Positions are often polarized and entrenched, based on perceived self-interest, as well as values, beliefs and views about what is right, good or just. Examples in the climate change discourse have included questions of who is to blame for climate change; what rights different countries should have to emit further greenhouse gases; what credit can be claimed for efforts to reduce emissions; and who should be responsible for paying for emission reductions.

Analysis of the dialogue on climate change is helpful. It allows us to understand how negotiating positions can be linked to cultures and belief systems as much as to vested interests. However, much existing discourse analysis views the debate as a self-contained process independent of the real power relations and processes of scientific discovery that help to shape it (Latour, 1993).

The current paper takes a slightly different approach. Its aim is to consider the potential for a future development in ethics that would support substantial greenhouse gas mitigation. It is concerned primarily with the ethics of the modern consumer culture: the way our ethics shape our attitudes to climate change; and the sources within our society of alternative systems of ethics. Ethics can be understood as an essential component of the narratives that we use to describe and communicate our society, our worldview, and our understanding of ourselves (MacIntyre, 1985; Taylor, 1989). Our ethical systems have evolved with developments in science, technology, institutions, and our view of ourselves.

The consumer culture sees increasing consumption as the main route to the good life. The values and ethics of this culture emerged from traditions that developed in Europe from the 16th century onwards. The early Enlightenment brought a commitment to progress, liberty, equality, the rational individual, and an instrumental view of nature (it viewed nature as existing for human use). The later Romantic Movement placed new emphasis on aesthetic appreciation, emotional individualism, creativity, self-expression, and the preservation of nature. These two traditions help shape the worldviews and self-images of most western individuals, and contribute to modern debates such as that over sustainable development.

Modern society is subject to a variety of unresolved tensions caused by internal inconsistencies in these traditions and the continual dialogue with other traditions. The last two centuries have seen continuing political struggle between the ideologies

E. Jochem et al. (eds.), Society, Behaviour, and Climate Change Mitigation, 157–168.

of liberty and equality, between the individual and the community, and between humanity and nature. Modern culture is committed to the freedom of individuals to pursue their own version of the good life, yet many people feel tied in to consumerism, which defines human well-being in material terms.

The next section will start by paying attention to some specific aspects of our current narratives, in particular what they tell us about ourselves, the society we live in, and the kind of life we should seek to lead. It will go on to consider the implications of the ethical elements of these narratives for climate mitigation. Finally, it will consider the way narratives evolve and the implications for climate change.

2. Cultural Narratives and Traditions

The study of narrative as a psychological, social and political force has gained increasing intellectual respectability with the work of Michel Foucault, Alasdair MacIntyre, Charles Taylor and others. Whereas Foucault was primarily concerned with the use of discourse, in particular writing, to establish and maintain power relationships, MacIntyre (1985) is interested in a more comprehensive kind of narrative. He describes a way of thinking about narrative as part of our cultural context or tradition, as something that we inhabit. Daniel Dennett (1993) has similarly described how comprehensively our assumptions about the world and ourselves and our ways of thinking derive, in one way or another, from pieces of narrative. We may receive them from our parents, teachers, preachers and friends, from books, plays and films. We may even make some of the up ourselves. They may take the form of fiction, history, science or philosophy. We weave them together in our minds until they become a more or less coherent story that guides our life and that (according to Dennett) provides us with a sense of self.

One of the most comprehensive surveys of narrative as both a reflection of existing culture and a force for cultural change is that by Joseph Campbell in *The Masks of God*. Campbell is concerned with what he calls mythology, a combination of narrative, symbol and ritual which amounts to something similar to MacIntyre's concept of a tradition. He observes four functions for mythology (Campbell, 1964) which I paraphrase:

1. to provide an understanding of our psychology
2. to provide a cosmology – an explanation of our position and role in the universe, in space and time
3. to give an intimation of something that we are part of beyond our tangible lives
4. to provide an ethical framework.

Charles Taylor has carried out a thorough exploration of the role of narrative in shaping our understanding of ourselves and the society we live in. He notes the co-evolution of: a) our notion of "the good"; b) our understanding of our self; c) our conception of society; and d) the narratives we inhabit (Taylor, 1989, p. 105). Such notions, understandings, conceptions and narratives are central to our responses to climate change.

Society in the process of globalization is host to a large number of cultures. This is partly a result of accumulation, but mostly due to our exceptional capacity in the last century for movement and communication. In addition, the embracing of pluralism is itself a feature of the "post-modern" tradition which has emerged in the last 30 years. Table 1 provides a suggested characterization, or perhaps a caricature, of the narratives within three current traditions: the "Humanist", "Modernist" and "Romantic" traditions. These three traditions emerged successively in Europe as intellectual and cultural movements following the Middle Ages.

Table 1: Illustrative characterization of concurrent traditions in Europe/North America

	Humanism	**Modern**	**Romantic**
Origins	14th century Italy: revival of classical Greek and Roman traditions	17th-18th century Netherlands and Britain: the Enlightenment	18th –19th century Europe building on Medieval romantic literature
Sphere of operation	The arts, architecture, philosophy, politics, science; concentrated in academic and intellectual circles.	Science, philosophy, business and government	Entertainment, advertising, private life
Cosmology	Values rational interpretation of empirical information. Anthropocentric view of the universe.	Big mechanistic universe, humans insignificant	Small, anthropocentric universe Apocalypse a result of the Modern tradition
Psychology	Multifaceted individual combining the rational, emotional and physical. Recognition of mixed motives and inconsistencies	Self-interested rational individuals	Emotional individual, social, expressive, Jungian
Attitude to nature	For enjoyment	Mastery, use	Idealized; the rural idyll; for enjoyment
The beyond	Transcendence through contemplation	None	Transcendence through aesthetic awareness
Ethics: Justice	Behaviour towards others guided by duties and virtues	Social contract; utilitarianism; individual rights; freedoms; equality	Individual rights and freedoms; equality
Ethics: Good life	Expression of the virtues; Aristotelian ideal of a well-balanced life allowing for rational contemplation.	Individually determined; material	Idealized loving couple or family setting.

It is particularly important for us to understand both Modernism and Romanticism. Not only do they shape the way we **think about responding** to climate change. They are also thought by many to be the traditions that underlie our system of industrial production and mass consumption (Corrigan, 1997); they may be in part the **causes** of climate change.

I have included the Humanist tradition in the table, partly because it remains an important part of culture in many European countries. It is also of interest because several contemporary philosophers have suggested that elements of ancient Greek philosophy, which the 15th century Renaissance movement celebrated, would be valuable in addressing some of the social and environmental difficulties that have been blamed on the modern and romantic traditions.

I must emphasize that some elements of these caricatures may be inaccurate and the table is intended mainly for illustrative purposes. Nor is the table by any means complete: each of these traditions contains many different specific narratives, and there are certainly traditions that are not included in the table. Indeed, there are probably more narratives than there are people in the world, since each person has several concurrent personal narratives.

It is rarely likely to be possible to allocate individuals entirely to specific traditions. Most people operate within several traditions depending on where they are, what they are doing and, most importantly, whom they are with. We are often expected to operate in the rational Modern tradition at work, but in the Romantic tradition at home, for example.

2.1. MODERNISM

We should probably concern ourselves most with the dominant tradition in the international policy and business community with which the IPCC is most closely associated. I would argue that this is Modernism, which is perhaps the most analysed and debated tradition of our time.

Modernism is essentially the Enlightenment tradition. It is usually traced back to philosophers such as Descartes, Locke and Hobbes, and to scientists such as Galileo and Newton. It brought a huge step forward from a feudal society ruled according to a "revealed" order. The Enlightenment brought the change in the European worldview and self-consciousness that allowed the Industrial Revolution to happen.

Cosmology

One of the key ideas within Modernism is that of linear development, rags to riches, life as a process of growth. The narrative can be found in the Big Bang theory of cosmogenesis, Darwin's theory of evolution, and the economic and technological history of society since the Industrial Revolution.

This Modernist narrative of progress can be found in much of the work of the international community. It is beautifully captured in the IPCC Special Report on Emission Scenarios, with its stories of economic growth and technological progress marching on through the 21st century. It is fundamental to the missions of international organizations such as the United Nations Development Programme and the Organization for Economic Co-operation and Development.

Efforts to counter this narrative have been mounted most famously in the writings of Thomas Malthus and the Club of Rome. Both have been largely discredited. One of the main arguments currently used against the narrative of progress is that economic growth is resulting in increasing environmental damage, especially climate change. Another source of doubt about the progress narrative is the widespread finding from surveys that material progress is not making people happier (Argyle, 1987; Inglehart, 2000).

Psychology

Modernism follows Descartes in viewing mind (the rational soul) and body as separate. Rationality is able to understand and operate on the material world. Philosophers in the Modern tradition have interpreted instinctive and emotional aspects of human consciousness and behaviour in a variety of ways. Thomas Hobbes viewed people as essentially selfish beings who move from desire to desire, and argued that they should enter into a social contract with the state to enforce order and prevent a destructive free-for-all. Jeremy Bentham in general saw reason as the means through which the material world should be organized to maximize happiness. However it is often unclear whether reason is serving the emotions, or the other way round. Freud can be seen as following the modernist tradition with his concepts of the animal *id*, the rational *ego* and the moral *superego*. Similarly, the Modern tradition includes Abraham Maslow's concept of people having a hierarchy of needs that they strive to satisfy, moving from physiological needs to those for belongingness and esteem, and ultimately to self actualization. The important feature that these descriptions have in common is that they see people as separate individuals (before they are social participants) somehow combining rationality with instinct.

Ethics

Modernist philosophers have developed a variety of ethical systems based on the principles of liberty and equality. The best known are:

- Bentham's concept of utilitarianism – the aim of achieving the highest level of happiness and the minimum of suffering for humanity at large;
- Kant's principle of universalisability – that individuals should only do what it would be right for everyone to do; and
- Rawls' idea of impartiality based on the "original position" – he proposed that the best system of justice is the one that we would choose in a kind of pre-incarnation state, where we had equal chance of becoming anyone, of any social or ethnic group, and with any position in society.

While Modernists have sought universal principles on which to base systems of justice, they have been very reluctant to embrace specific visions of the Good Life. It is not up to the Modernist society to tell people how to live. One of our most important freedoms is that of determining our own conception of the good life. The social ideals in Modernism are ideals of structure and process – the market, capitalism, socialism, and above all, progress.

Of the three major Modernist ethical systems, Bentham's (1789) concept of utility is perhaps the most important for our purposes, because it underlies the economic analysis that is usually applied to thinking about climate change. Utilitarianism has been heavily debated by philosophers and several aspects are now rejected by most. He asserted that society is no more than the sum of its members, and that the greatest good for society is the pursuit of the greatest happiness and the minimum of pain for its members. He believed that it should be possible to calculate the utility of the members of society and to evaluate legal and political courses of action based on the effects they would have on this quantity. While this concept clearly remains strong in welfare economics, few philosophers or sociologists find it convincing. John Stuart Mill, who called himself a utilitarian, could not accept Bentham's view that all forms of pleasure could be reduced to a single metric, and embraced an idea of "higher" pleasures reminiscent of Plato's view of rational contemplation as the highest form of happiness.

Another widespread narrative in the Modernist tradition is that of the nation state made up of free and equal individuals who are the bearers of individual rights. The idea that individuals have rights has its roots in the English Magna Carta and was developed especially by the English 17th century philosopher John Locke. It played an important role in the English, French and American revolutions and is expressed in the English and American Bills of Rights. The UN's 1948 Universal Declaration of Human Rights has almost universal acceptance. Perhaps the most fundamental of our rights is that to liberty: liberty from oppression, coercion, suffering, unjust punishment; and liberty of thought, speech, action, to own property, and to pursue our own conception of the good life. The nation state was fundamental in Locke's thinking and remains, in the UN Declaration, as the counterpart to the individual in the social contract. Every individual has the right to be a citizen of a nation and the nation has the obligation to protect the rights of its citizens.

Within the Modern tradition, with its narratives of progress, individual rights, liberty and equality, there are a number of disputes about which narrative is the most important. Should we be willing to forego some equality for a bit more liberty or progress? How should we trade off my right to what I have earned against your right to meet your basic needs? These two quarrels have been played out between right and left, West and East, over much of the last century. They are a central part of the climate change debate as described by Thompson and Rayner (1998).

Ultimately, the ethical systems of the Modernist tradition rest on axioms, which are asserted or postulated by their proponents as universal principles. However, there is no basis within the Modernist tradition for choosing one principle over another. What seems self-evident to the various proponents often appears to be utter nonsense to others. What starts as nonsense becomes highly dangerous when carried to extremes, and critics of narratives in the Modern tradition can point to numerous inhumanities that have been perpetrated in the last two centuries in the name of one universal principle or other.

Relationship with nature

According to the Modern narrative, nature is there to be developed or used. At an early stage, Modern thinkers began to establish clear boundaries between humanity and nature (Latour, 1993). Society and the natural world are taken to be completely different phenomena and they are studied and described by different professions. Nature is there for society to use. In taking control of nature and making it economically productive, we improve it (Locke, 1698). Modernism centres on human beings as the locus of agency in the universe. Some older traditions viewed nature as important in itself because they saw God, Brahman or the Fates as immanent in it. To the extent that Modernism embraces a concept of God, it is a being separate from creation. Nature is created by God to be used by human beings.

2.2 ROMANTICISM

While Romanticism strictly refers to a specific movement in the arts at the end of the 18th century and the beginning of the 19th, I am using it here to describe an ongoing tradition of emotionalism. I will distinguish in the following between the historical Romantic movement and the current Romantic tradition.

The Modernist tradition dominates our society in many ways, but the Romantic tradition is also important – in fact it is inseparable from Modernism. The consumption that drives economic growth is fuelled by Romantic narratives about the good life, which centre on enjoyment of consumption as well as relationships with other people and with nature. Much consumption is motivated by the desire to achieve an ideal of romantic love, which is a narrative within the Romantic, not the Modernist tradition. Romanticism is also important as a parent of the environmental movement.

The Romantic movement was a late 18th century reaction against the Modernist ideal of progress, which appeared to be advancing the interests and worldview of one group of people at the expense of others. It also seemed to them to value industrial development and money over nature and reason over emotion. The movement spread, and became a tradition, through the rise of the novel. Romantic novels drew on older roots, including medieval literature such as the Arcturian legends. The original of the Romantic ideal of the individual can perhaps be found in Wolfram von Eschenbach's Parzival – the knight who found the Holy Grail and became the Grail King, by following his own heart. Perhaps even more important is the ideal of romantic love, which can be found in Gottfried's Tristan and Isolde (Campbell, 1968).

It is easy to recognize the continuing Romantic tradition in our society. It is present especially in literature and the cinema with the emphasis on being true to one's own passions and on romantic love, and with the tendency to anthropomorphic representations of animals, sanitized views of wilderness, and idealized cameos of family life.

Cosmology and psychology are closely linked in the Romantic tradition. It tends to treat emotions and aesthetics such as love and beauty as fundamental principles that underlie the universe. In some ways this approach is related to Plato's idea of the Good. Such concepts can also be found in astrology and alchemy. Within this worldview, human psychology is an expression of those principles. The psychologist Carl Jung fits well within the Romantic tradition.

The Romantic movement created a new ethic that celebrated aesthetic appreciation for its own sake. The Romantics made it somehow noble to appreciate a beautiful view, good food and fine clothes (Campbell, 1983).

2.3. HUMANISM

Humanism was the original movement within the Renaissance, beginning in the 14th century in Italy. It involved a new openness to learning, and a revival of ancient Greek and Roman literature and artistic styles and of secular learning and discourse. The movement was focused on understanding human nature and the good life. Its heritage can be seen in approaches of the modern social sciences and in a substantial part of the middle class intellectual culture of Europe and North America.

The Humanist conception of human nature is richer and more holistic than those of the other two traditions reviewed here. People are understood to have internal contradictions, and human welfare is understood to depend on multiple, incommensurable conditions. Hence, the Humanist conception of the good life includes the development of physical, mental and emotional capabilities. This idea was propounded by Aristotle in his *Nicomachean Ethics* and can still be seen in the late 20th century ideal of the "new" man or woman, whose culture values healthy food and regular exercise as well as emotional and intellectual authenticity.

It is the Humanist ethical system that is probably of most interest in considering climate mitigation, and sustainable development more generally. Whereas the Modernist tradition has tended to seek simple, universal principles on which to base its ethics, Humanists emphasize the need for a judicious mix of the virtues. Aristotle described an essential part of moral virtue as a disposition to choose the mean between excess and deficiency in worldly pleasures. He saw the good life as a life of happiness based on the exercise of the rational capacity and moral virtue, but also requiring material means and relationships with family and friends.

European Humanism has resonance in other, older traditions from other parts of the world. In particular, similar conceptions of a good life based on development and expression of the virtues can be found in Taoism, Hinduism (especially the Bhagavad Gita), and Buddhism.

2.4. OTHER TRADITIONS

We might recognize an attempt in the dialogue on sustainable development to create a new tradition. If so, this tradition is rather poorly defined. Indeed, it is probably fairer to say that sustainable development is an attempt to bring together several divergent traditions. It tries to combine values from Modernism, Romanticism, and elsewhere. Nevertheless, there are some core narratives. There is broad agreement within the growing literature on sustainable development on two key ideas. One is that that human beings play a significant role in our planet's ecology, and that a change in direction is needed if we to are to avert environmental destruction. The other is that too many of the world's people are not meeting their basic needs, and that a change in the direction of our development efforts is needed to enable them to do so.

European and American society retain a strong Judaeo-Christian tradition, which is very different to the others illustrated here, partly because it predates them. Nevertheless it remains, along with other classical and more recent traditions, as a significant source of counter-narratives opposing the tendencies of Modernism and Romanticism. These counter-narratives may provide useful ideas that could be incorporated into the Sustainability tradition.

It has been mentioned that Buddhism, Hinduism and Taoism have some common ground with European Humanism, in particular with their emphasis on the virtues and the good life. The core value in the more mystical reaches of the religious traditions is transcendence, becoming closer to the divine or to some kind of universal consciousness. Transcendence involves non-attachment to self, desires and property. Nevertheless, some of these narratives value the generation of personal wealth, provided that it is gained through a life of virtue and is shared with the needy. On the whole, they counsel moderation, seeing greed, selfishness and acquisitiveness as vices.

One of the most potent counter-narratives to have emerged recently is that of Marxism. In fact, Marx accepted a large part of the Modernist narrative, including its emphasis on progress, liberty and equality. However, he did not share its conception of the nature of human beings as autonomous individuals. He rather saw the individual self as something that was produced and shaped by the social context, which was driven by the top-down process of evolution of social and economic forms.

3. Views on climate change from the traditions

So how do the narratives we inhabit shape our views of climate change and our potential responses to it? In particular, what needs to change, and how should the effort to change be distributed?

Within the Modern tradition, and especially in the utilitarian ethical system adopted in economics, climate change is generally viewed as an optimization problem that might require a shift from current patterns of production and consumption. It could be a more serious problem if it interferes with the core narrative of progress, but calls to action to mitigate climate change seem to pose a greater threat because some of them seem to question the heart of the Modern tradition.

The debate within Modernism is centred on how greenhouse gas emissions might be reduced to a satisfactory level, consistent with the principles of liberty, equity and continuing progress. This debate is analysed in detail by Thompson and Rayner (1998).

For utilitarians, what matters is the impact of climate change on human happiness and suffering, represented in a benefit-cost framework. This framework supports solutions which optimize benefit minus cost regardless of the distributional impacts, unless any inequity in the distribution of costs and benefits can be incorporated into the framework as a cost in itself.

Rawls' concept of the original position suggests that we should aim for equal levels of individual cost in choosing mitigation targets, although his practical ethics point to negotiated emission limits.

The Kantian system provides a very different way of thinking about our responses to climate change. It suggests that it is only right for me to emit a certain level of greenhouse gases if that level could be universalized. This system points towards global equity in per capita emission limits.

From the Romantic tradition, climate change appears to be a rather different kind of problem. It is, in fact, a problem caused by the Modernist tradition itself. The tendency is to blame Modernism as represented by business and industry, which should change its production methods so that it does not damage nature. Narratives in the Romantic tradition tend not to recognize the role of Romanticism in shaping the consumption patterns for which industry produces.

Viewed from the European Humanist tradition (and also in slightly different ways from Buddhism, Taoism and Hinduism), climate change appears to be a problem resulting from a failure to nurture virtues such as humility and continence, and to control vices such as greed, pride and acquisitiveness. At root, this can be seen as a failure in Modernism to understand human nature, and the nature of the Good Life. Too much emphasis is placed on material consumption and not enough on developing family relationships, communities, civic involvement, and opportunities for learning and contemplation.

4. Where next?

Writers such as Alasdair MacIntyre and Arran Gare see little hope within the Modernist tradition for solving the problems of our time, because Modernism is itself part of those problems. MacIntyre's view is particularly negative and he suggests that European civilization is in the process of collapse because it has lost its view of the Good Life. He believes that a return to Humanism is necessary. However, there are clearly great values within the Modernist tradition and it is perhaps more helpful to view Modernism as one stage in the development of human society. We might agree with Marx and Schumpeter in thinking that liberal capitalism has been a valuable step forward from feudalism but that it should not be the final resting state for society.

Some social scientists believe that Modernism is giving way to a new postmodern tradition. Others are dismissive of postmodernism, seeing it as a passing phase. The word "postmodern" is used in many different ways, and mostly refers to forms of architecture, art, and literature. Postmodern architecture and art involves an assemblage of forms from different periods and traditions. If postmodernism is a new tradition, it is one that embraces multiple discourses, that sees all knowledge as relative, and that (for some adherents at least) sees reality as a social construction.

While postmodernists are intensely critical of the Modern and Romantic traditions, Gare (1995) views postmodernism as holding out little hope for the environment. In allowing for all narratives to be equally valid, postmodernists also make them all equally invalid. Latour (1993) similarly dismisses postmodernism as a tradition locked into the confines of discourse and failing to engage with the world of nature or society.

If postmodernism does not contain the answers, it may help to make sure the right questions begin to be asked. By opening up the boundaries between fact and fiction, and between academic disciplines, and encouraging social scientists at least to rethink their methods, it could be a way-station to the development of new traditions that are more inclusive of environmental concerns.

The IPCC needs to engage with the social sciences as they struggle to recreate themselves. Future cultural developments are not likely to be under anybody's control, but there will be developments and they will have implications for climate change. There is an urgent need for the IPCC to devote more attention to the processes underlying the evolution of cultural traditions and narratives, and to make greater efforts to bring natural scientists and social scientists together.

If the natural scientists do not learn to analyse themselves and their assumptions in the way social scientists are beginning to, they will not be able to contribute effectively to the new, emerging tradition. This would be unfortunate because that tradition will not necessarily incorporate a strong concern for the natural world. Inglehart (1990, 2000) uses extensive survey evidence to argue that younger people are developing increasingly "postmaterialist" values. This postmaterialism involves placing much more weight on developing relationships and social networks, rather than on developing specific professions and on material consumption. But it offers little hope for climate change: postmaterialist values are important drivers for the current growth in energy-intensive travel.

5. References

Bentham, J., 1789 (1987 publication): An Introduction to the Principles of Morals and Legislation. In: *Utilitarianism and Other Essays*, Penguin, London.

Campbell, C., 1983: Romanticism and the consumer ethic: intimations of a Weber-style thesis. *Sociological Analysis*, 44(4): pp. 279-96.

Campbell, J., 1964: *The Masks of God, Volume 3 of 4, Occidental Mythology*, Penguin, New York.

Campbell, J., 1968: *The Masks of God, Volume 4 of 4, Creative Mythology*, Penguin, New York.

Corrigan, P., 1997: *The Sociology of Consumption*, Sage Publications, London.

Dennet, D., 1993: *Consciousness Explained*, Penguin, London.

Gare, A.E., 1995: *Postmodernism and the Environmental Crisis*, Routledge, London.

Inglehart, R., 1990: *Culture Shift in Advanced Industrial Society*, Princeton University Press, Princeton, NJ.

Inglehart. R., 2000: Globalization and Postmodern Values, *The Washington Quarterly*, Winter 2000, pp. 215-228.

Latour, B., 1993: *We Have Never Been Modern*, Prentice Hall, Harlow, England.

Locke, J., 1698 (1982 publication): *Second Treatise of Government*, Harland Davidson, Inc., Wheeler, IL.

MacIntyre, A., 1985: *After Virtue*, Second edition, Duckworth, London.

Maslow, A., 1954: *Motivation and Personality,* Harper and Row, New York.

Ney, S., M. Thompson 2000: *Cultural discourses in the global climate change debate*, in Society, Behaviour and Climate Change Mitigation. E. Jochem, D. Bouille, J. Sathaye (eds.), Kluwer, Dordrecht, The Netherlands.

Taylor, C., 1989: *Sources of the Self: The Making of the Modern Identity*, Cambridge University Press.

Thompson, M., 2000: *Consumption, motivation and choice across scale: consequences for selected target groups*. In: Society, Behaviour and Climate Change Mitigation. E. Jochem, D. Bouille, J. Sathaye (eds.), Kluwer, Dordrecht, The Netherlands.

Gender-Specific Patterns of Poverty and (Over-)Consumption in Developing and Developed Countries

Minu Hemmati

1. Introduction

Much as other psychological, social and cultural aspects of consumer behaviour have not been the focus of approaches taken to addressing climate change, gender issues have rarely been addressed as part of the discussions. Human behaviour cannot be sufficiently understood on the basis of a purely rational or technology-oriented analysis of 'external' conditions such as available economic and natural resources and the related choices and barriers. Behavioural patterns are the result of complex interactions of 'internal' and 'external' conditions; of cognitive and emotional factors; perceptions; evaluations; social pressures; cultural images; physical environments and economic conditions. People's choices are based on needs of immediate survival as well as inspired by the need for social acceptance and images of desired identities.

In its study of approaches to consumption, the Organization for Economic Co-operation and Development (OECD) pointed out that "the discussion of 'consumption patterns' has tended to remain at a relatively superficial level because it is difficult to piece together in a comprehensive framework all the influences that shape what and how societies consume". The report goes on to say: "developing effective and efficient policies to encourage behaviour change requires a better understanding of the various facets of consumption patterns and their environmental impacts, in order to pinpoint; [1] where consumption patterns are susceptible to change [2] where in the product lifecycle is the best point, or points for intervention [3] which actors in the network, including government, are likely to be the most effective agents of change".

---

[1] For an overview of international agreements on women and sustainable development issues which came out of the cycle of UN Summits and Conferences in the 1990s, see Hemmati et al., 2000.

[2] The other fundamental axiom is that people construct their own realities. Considered in conjunction with the pervasiveness of social influence, this also means that people construct their own realities under the influence of their social (and physical, as it represents history, culture and society) environment.

[3] As a note on the side, we should keep in mind that through the cited evidence we do not know what women are 'really' like, what their capabilities are or could be. All we know is that all men and women have been growing up in societies where gender roles differ and gender identities differ. The 'ultimate experiment' in this matter would comprise designing a society with reverse gender roles or a society with no gender differences at all. As we cannot conduct such an experiment, we will need, inter alia, more research in genetics to clarify hereditary sources of characteristics, and further developments of societies towards gender equity, to get a clearer idea of women's and men's capabilities.

E. Jochem et al. (eds.), Society, Behaviour, and Climate Change Mitigation, 169–189.

Consumption is an inevitable part of our lives but what and how we consume is socially constructed, the more so the more consumption choices we have. All production of what we consume requires energy inputs, which are mostly produced by burning fossil fuels for electricity generation or transport. Energy consumption is one of the major causes of greenhouse gas emissions, hence reducing energy consumption would enable us to reduce greenhouse gas emissions. Therefore, understanding consumption behaviour and developing strategies to influence it is one of the building blocks of climate change mitigation.

This paper aims to discuss gender as a social category and an important determinant of human behaviour. The paper focuses on the consumption behaviour of women and men in several areas, mostly guided by what is relevant to mitigating greenhouse gas emissions but also by what research is available.

It is crucial to understand that gender is an important social category, influencing almost every aspect of people's lives. As Costa (1994, viii) points out, "because gender is pervasive, intricate and interwoven with virtually all aspects of human behaviour, further study is necessary if we are to understand more fully this important dimension of society and individual behaviour".

However, gender is certainly not the only social category impacting on the thinking, feeling and behaviour of people, social structures and power relations. Other social categories such as age, income level, class, ethnic group membership, religion, etc. are also powerful determinants of people's roles and identities.

"A gender perspective helps redefine the public from one amorphous blob of humanity with average needs, to publics with diverse experiences, needs and aspirations" (Figueroa 1998). Gender as a crosscutting issue adds value to considering mitigation assessment. An integration of gender provides a structural dimension, as gender equality is key to any paradigm for sustainable development. As Charkiewicz (1998, p. 3) points out, "The attention to gender differences in project and policy formulation improves the effectiveness, equity and sustainable outcomes of development. Gender equality has become an intrinsic, normative goal of sustainable development and practically every chapter of Agenda 21 acknowledges the contribution of women to sustainable development" (see footnote 1).

2. Concepts

For the sake of shortening this paper, it does not provide an overview of theoretical approaches to studying gender. Numerous and fruitful approaches of feminist anthropology, theory of science, sociology and social psychology have been developed over the last decades. Many of them can be used as a basis for studying consumption behaviour. The approach of this paper is a social psychological one. A few of the relevant concepts used in this approach are briefly presented below.

Whereas the term *'sex'* refers to biological differences between women and men, the term *'gender'* refers to social differences. In societies, women and men fulfil different gender roles, and gender stereotypes describe 'typical' or 'ideal' sets of characteristics of women and men. Based on the societal 'images' of what is feminine and what is masculine, individuals develop gender-specific identities. As a reflection of roles, stereotypes and identities, women and men differ with regard to motivation and behaviour.

One of the two 'fundamental axioms of social psychology' (Smith & Mackie 2000, p. 16ff) comes into play here: The pervasiveness of social influence. The term refers to the axiom that other people influence virtually all our thoughts, feelings, and behaviour, whether those others are physically present or not (see footnote 2).

"*Social norms* describe generally accepted ways of thinking, feeling, or behaving that people in a society agree on and endorse as right and appropriate" (Smith and Mackie 2000, p. 594). There are different social norms for different social groups in societies, such as those describing women's and men's appropriate ways to think, feel and behave. Gender specific social norms define roles for women and men to fulfil. Social acceptance is one of the consequences of fulfilling gender roles, and social sanctions are likely to occur when people do not comply with gender roles - with some flexibility determined by subgroup membership and individual interpretation.

Stereotypes are cognitive representations or impressions of a social group that people form by associating particular characteristics and emotions with the group (e.g. Mackie and Hamilton, 1993; Zanna and Olson, 1994). Many different kinds of characteristics are included in stereotypes, which can be positive or negative. Some stereotypes accurately reflect actual differences between groups, though in exaggerated form. Other stereotypes are completely inaccurate.

Gender stereotypes have been widespread in every society throughout history. Although they have differed across times and cultures, there are significant commonalities, such as: Women are supposed to be soft-hearted, caring, submissive, dependent, affectionate, anxious, emotional, sensitive, sentimental, with a sensitive and emotional leadership style. Men are supposed to be aggressive, independent, strong, tough, autocratic, with a dominant leadership style.

Much research has been directed to the question of how much these stereotypes are reflected in the *actual behaviour*, thinking and feeling of women and men - or, how much stereotypes reflect actual differences. Metaanalyses have shown that men are overall more aggressive than women, particularly regarding physical rather than psychological aggression and in situations where aggression may be dangerous to oneself. Women have been shown to be more influenceable than men, particularly when influence is exerted by a group rather than through persuasive messages. Also, the difference is larger when a topic is regarded as 'masculine'. Women are more non-verbally expressive and more non-verbally sensitive than men. As leaders, women are more democratic and men are more autocratic; however, this difference is larger in laboratory studies than of studies of leadership in real, ongoing organizational processes (Eagly 1987, Eagly & Johnson 1990) (see footnote 3).

Growing up in a society where gender roles and stereotypes prevail and education continues to be gender-specific, individuals develop *gender-specific identities*. What women and men think and feel about themselves is partly determined by what they have learned they should be like. People compare themselves against these social standards and develop goals of personal development in accordance (or in contradiction) with those standards. Thus, a gendered social environment is reflected not only in attitudes and in overt behaviour, but also in the self-concept and in personal goals ('ideal self').

Women's and men's motivation differs, or, more precisely, the ways in which women and men pursue certain shared goals differ. For example, women and men share the need for social acceptance and hence a basic motive to be accepted by their peer groups. However, based on their gender-specific roles and identities, they differ with regard to which behaviour will serve the goal of being accepted.

Regarding the context of consumption, we know, for example, that women first address the needs of their families, particularly their children, whereas men are more likely to spend resources for their individual needs (see below). Both sets of priorities are gender-related and socially co-determined. Social status and social acceptance, for men and women, is based on fulfilling their respective roles and being 'female' and 'male' in the prescribed manners.

Thus, given their roles and identities as sensitive and responsible care-givers, women should be more highly motivated to ensure that the needs of their children, of elderly, disabled and ill persons in their care, are being met. Fulfilling this role, women gain social acceptance and status as women. Similarly, as men's social status is based on, for example, being independent, they should be more highly motivated to exhibit their independence by caring visibly for their individual needs. Men's status is also more closely related than women's to economic well-being; consequently, exhibiting wealth, for example through luxurious consumption, is a means of demonstrating and gaining status for men.

Another important aspect, linked to social status, is what determines attractiveness of women and men as potential partners for members of the respective opposite sex. Whereas women's attractiveness is rather determined by physical attractiveness (as defined by a cultural standard), youth and friendliness, men's attractiveness is more reliant on income level, intelligence, dominance / independence and a sense of humour. Many consumption choices can be co-determined by what makes people attractive - women investing in fashion, cosmetics and make-up, for example (see 3.3), and men in prestigious cars (see 3.5). Both have implications for energy consumption.

Consumption patterns in developed countries vs. developing countries differ significantly. However, there are overlaps, particularly regarding elite minorities in developing countries and people living in poverty in developed countries. What is said here about Northern and Southern issues related to consumer behaviour in the North vs. the South refers to the *'Global North'* and the *'Global South'* rather than the developing and developed countries as such. Evidence indicates that with globalization a host of consumption options have opened up for people in developing countries (HDR, 1998). Market research has identified 'global elites' and 'global middle classes' that follow the same consumption styles in different parts of the world, showing preferences for global brands (ibid.).

In a situation where survival is relatively secure and people live with a certain amount of resources, comfort and security, other motives and interests come into play than if they need to struggle daily to meet their basic needs. Consequently, gender-specific consumer behaviour differs significantly between North and South. In the Global South, we are dealing with poverty and survival. In the Global North, we are dealing with an expanding 'consumer culture', largely influenced by the Anglo-Saxon market system and its values. In a nutshell, environmentally damaging behaviour in the North is mostly caused by over-consumption, and in the South by lack of better choices due to poverty. In both cases, gender is an important category. Women make most of the day-to-day consumption choices for themselves and their families. Women are the major suppliers of food (shopping, cooking). And women are the majority among the world's poor.

In the North, women are the ones managing the day-to-day household and its budget, and are being confronted with increasing demands by their children, influenced by consumer culture, and the need to make informed and healthy choices. In the South, women have the same responsibilities for managing the households, but they are faced with incomparable levels of poverty and trying to accommodate the family's needs on the basis of resources that are often virtually non-existing except for their own work, time and energy. Globally, women are the ones who have proven to be less concerned about themselves as individuals than men, putting the needs of their families first, and making responsible choices for them. In some cases, it can be shown that women make the more 'sustainable' choices (see below). There is evidence that this is due to lack of resources, to less interest in fulfilling individual consumption desires and/or to higher environmental awareness – although the existing research does not provide a fully consistent picture. One of the questions for future research and metaanalysis will be to clarify further under which conditions and in which subgroups these differences account for behavioural differences between women and men.

In a global analysis of gender-specific consumption patterns, we have to keep in mind the inter-relatedness of the two mentioned relevant categories - the gender category and the category of access and control over resources. Both categories are powerful determinants of behaviour.

3. Issues & Research

This paper does not aim to give a full overview of existing research in the area of gender, (over-) consumption and poverty as it relates to the mitigation of climate change. Based on existing research, several issues, trends and respective findings - mostly regarding behavioural differences - are being presented below to provide a basis for discussion about how those behaviours might be analysed and eventually changed.

3.1. POVERTY AND (OVER-)CONSUMPTION

The Human Development Report's (HDR 1998) approach to consumption is based on an interesting definition: "From a people's perspective consumption is a means to human development. Its significance lies in enlarging people's capabilities to live long and to live well. Consumption opens opportunities without which a person would be left in human poverty" (ibid., p. 38).

In its definition of consumption, the HDR includes non-material consumption, i.e., social security, health, education, childcare and transport. It also includes consumption that lies outside the monetized economy, i.e., goods and services supplied through unpaid work, especially by women.

The HDR 1998 brings to light that despite a dramatic surge in consumption more than one billion people lack the opportunity to consume in ways that would satisfy their most basic needs. The report addresses two issues of consumption, i.e., the poverty and inequality nexus and the poverty and environment degradation nexus.

With regard to the poverty and inequality nexus, the HDR informs us that, within a global context, there are widespread disparities regarding what people can consume. For example, 20% of the highest income countries account for 86% of total private consumption, while the poorest 20% account for only 1.3%. The richest fifth consume 58% of total energy, while the poorest fifth consume less than 4%. The HDR 1998 (p. 51) points out that of the 4.4 billion people in the developing world, nearly three-fifths lack access to sanitation, a third have no access to clean water, a quarter do not have adequate housing and a fifth have no access to modern health services of any kind. Shortfalls of essentials are not just a problem for poor countries; in developed countries many individuals cannot meet their basic needs (Global South). The central point that the HDR (1998 p. iii) makes is that the poorest have been left out of the consumption explosion. For more than one billion of the world's poor, increased consumption is a vital necessity and a basic right – "a right to freedom from poverty and want".

Globally, *poverty has a woman's face*, i.e., of the 1.3 billion people living in poverty, 70% are women (HDR, 1995, Haq 1997). Women also constitute two-thirds of the world's illiterates (ibid.). Global statistics place them behind men in relation to health, education, nutrition levels, political participation, legal rights, equal pay for equal work, among many other things (Bruce & Dwyer, 1988; Haq, 1997; HDR, 1995).

Women also have less access to credit in developing countries. To give some examples (HDR, 1995), in Latin America and the Caribbean only 7-11% of beneficiaries of credit programmes are women. A study of 38 branches of major banks in India found that only 11% of the borrowers were women. In Zaire women made up only

14% of borrowers from commercial banks. Most banks in developing countries require that borrowers be wage earners or property owners who can provide acceptable collateral. In most countries such borrowers are men. Lack of monetary income, land or other real property in their own name, and limited education, keep women from accessing credit.

As women are poorer than men in most societies, they are the ones who suffer the lack of basic necessities (Mananzan, 1999). Moreover, the lower incomes of women also deny them basic rights as consumers. The principal rights of the consumer – access to essential goods, choice, safety, information, representation, redress, consumer education and a healthy environment are least attainable by poor women (Wells & Sim, 1987). Poor women who are illiterate are especially vulnerable to unethical marketing practices such as higher prices and fraudulent services.

At the same time, there has been an enormous increase in consumption levels world wide, mostly among the privileged. World consumption expenditures, private and public, have expanded at an unprecedented pace, doubling in the last 25 years to reach $ 24 trillion in 1998 (HDR, 1998). However, apart from an increase in goods and services used to meet people's needs, there are other factors that have motivated excess and conspicuous consumption. Spending is often motivated by pressures to match social status and standards set by society. Household debts, especially consumer credit, are increasing while household savings are falling in many developed and developing countries (ibid.). As CI ROAP (1997) points out, the 'keeping up with the Joneses' syndrome is especially apparent in societies with significant gaps in income.

Conspicuous consumption can have adverse effects on poor people and gender relations when consumer aspirations are motivated by social pressures. Increased spending whereby households aim to emulate the lifestyles of wealthy people can crowd out essentials such as food, education, and health care (HDR, 1998). This can have a discriminatory effect on girls since evidence shows that they receive less food, education and health care in developing countries vis-à-vis boys (DasGupta, 1993; Haq 1997; Dreze & Sen, 1995).

However, the adverse effects of conspicuous consumption on gender relations need to be investigated more thoroughly. The issue of globalization is also linked to excess consumption. The integration of the global consumer market has brought changes in consumption patterns that are particularly apparent in Asia and Latin America (Robinson & Goodman, 1996). The issue of globalization's impact on consumer behaviour and the gendered implications need further investigation.

3.2. CONSUMPTION AND GENDER

Women represent the largest group of consumers or shoppers world-wide (Beckmann, 1997; Mananzan, 1998). According to Wells & Gaik Sim (1987): "All of us are consumers, women and men of all nationalities, all social classes and groups. But women are the largest group of consumers, buying for others as well as for themselves." Grunert-Beckmann (1991, p. 625) says: "Women represent the largest group of consumers. They take part in the consumption cycle – choosing, buying, using and disposing – both for themselves and for others. The second task is often the dominant one; because a woman is responsible for most of the shopping does not necessarily imply that she uses what is bought. On the contrary, as the family manager she often buys what suits her husband and children rather than herself."

Globally, women are the ones who make the purchasing choices of daily items. Women are mainly responsible for the everyday shopping of their households. In the UK, women make over 80% of consumer decisions (Mawle, 1996; Vajpayi, 1996). According to Costa (1993), in the US women are often responsible for consumption activities – shopping, preparing items for consumption, gift buying and disposal of used items. She points out that in some cases American men are typically more responsible for the purchase of certain types of goods than are women. Firat (1994) foresees that the deconstruction of gender roles in post-modern culture is likely to create radical transformations in both consumer behaviour and marketing. However, the deconstruction of gender roles in post-modern culture is viewed quite sceptically by some scholars, who argue that little has changed from modernity to post-modernity (MacDonald 1995, Costa 1994).

3.3. CONSUMPTION CHOICES BY WOMEN AND MEN

There is evidence from developing countries of different income allocation priorities amongst men and women. Evidence indicates that men tend to a much larger extent to spend their earnings on personal consumption (for example, cigarettes, liquor etc). On the other hand, women's earnings are prioritized on their children and family needs. As Bruce & Dwyer (1988, p. 5) argue, "At issue is not simply the ways in which women's income is used, but the degree to which men and women differ in taking personal spending money from their earnings. Though the specifics of women's consumption responsibilities vary (in Africa and across the world) it is quite commonly found that gender ideologies support the notion that men have a right to personal spending money which they are perceived to need or deserve and that women's income is for collective purposes."

It is also important to examine the nature of major consumption decisions that men and women make in the family as this is linked to the use and control of resources – for example, allocation of food, clothes, medicines, and decisions on education etc. There is compelling evidence from developing countries that resources under the control of women are more likely to be devoted to children and are likely to contribute significantly to the well-being of the family (Bruce & Dwyer, 1988; Moore, 1994, Thomas, 1990). For example, Thomas (1990) found that income that was controlled by Brazilian women increased the health and survival chances of their children to a level twenty times greater compared with income that was controlled by the father (Moore, 1994).

A rather reverse example of gendered consumption in developed countries is fashion. Fashion is mostly a feminine area of consumption, although increased marketing of fashionable clothes and cosmetics targeting men has begun to change this dichotomic picture. Women buy far more fashionable clothes than men and tend to spend a far greater percentage on their appearance that men do. This is consistent with women's roles and identities. For women, to be socially accepted and perceived as desirable by the opposite sex, it is important to look attractive and beautiful. In modern Western societies, attractiveness and beauty is linked to being (or looking) young, slim, healthy, and 'well-maintained'. Fashion, as well as cosmetics and make-up, helps to portray these attributes. A fashionably dressed woman will, on average, be perceived as attractive. The enormous increase of consumption levels referred to above has also occurred in the area of fashion. Fashion cycles have also become shorter and new collections come out now every three months. There has been an increase in lower price market sales as well as in extremely luxurious goods. Production has been moved to low income countries and a lot of child labour and sweat shop labour is involved. Fashion consumption in developed countries is in many cases an example of over-consumption, and also in many cases has adverse impacts on developing countries – regarding both production and environmental impacts [4].

3.4. ENERGY CONSUMPTION

Lack of access to fuel (firewood and alternative sources) in developing countries causes people to cut down wood and forests in an unsustainable way and contributes to deforestation. The people who lack access to resources and are at the same time responsible for food and safe water supply are women. It is well known that women mainly bear the burden of providing biomass fuels for domestic use (eg Batliwala & Reddy 1996). Lack of access to modern cooking and heating technologies forces women into often unsustainable use of natural resources. This lack also largely contributes to women's disadvantages in development, taking up their time and energy. Women have significantly less of their own human resources available for education and income-generating activities.

Even access to technologies which reduce the work load does not necessarily benefit women. For example, mechanization of farming and introducing tractors for ploughing has had gender biased results in many African countries. While the men took on the task of riding the tractors and increasing their work efficiency, women have remained responsible for the weeding and harvesting, manual tasks predominantly performed by women. "Thus, mechanisation has served not only to deforest large areas of Africa (thereby affecting its carbon sequestration capacity) but also tremendously increased the agricultural workload for women" (Raban Chanda, review communication).

---

4 A recent protest letter from the Indian Minister Maneka Gandhi to the fashion designer Calvin Klein can illustrate this point. Gandhi claims that by using snake skin in his latest designs, Klein has contributed to environmental damage and food insecurity in India. Gandhi continues by saying that because of the increased demand for snake skin in the fashion market, more snakes than ever before have been caught in India and are being (illegally) exported. This has caused an increase in the rat population which in turn has led to the loss of up to 50 % of the rice crop yield in India (DER SPIEGEL; 28/02/2000, p 260).

Cross-country data show the linkages between energy consumption and the distribution of income. Total per capita energy consumption increases with the per capita GDP. The mix of energy carriers varies with income and its distribution (Leach 1992). In particular, reliance on biomass is greater among countries with lower incomes, among countries with more unequal income distributions, and among countries with relatively small urban populations. The end-uses of human energy in villages show that the inhabitants, particularly women and children, face enormous burdens, e.g. gathering fuelwood and fetching water. There are also serious health implications arising from rural energy consumption patterns which imply over-use of fuelwood and other biomass energy (eg Batliwala 1982, 1987). Health problems relating to energy consumption affects women and girls in particular. Because they are responsible for the preparation of food for their families, women and girls are much more exposed to indoor air pollution, resulting in diseases of the respiratory system and the eyes.

The other side of the 'inequity coin' in relation to energy consumption is over-consumption of energy in the Global North. The production and use of luxurious goods such as unnecessary, energy-intensive electric household appliances (alarm systems, communication and entertainment systems, jacuzzis, saunas, dryers, etc.) requires extremely high levels of energy use. Gender differences in access to these goods usually only become apparent when differentiating within households by income and decisions about income use. Transport is one of few issues which have attracted more research in recent years (see below); with regard to many of these goods, research is sketchy at best.

3.5. TRANSPORT

The growing literature on women and transport has clearly shown that women tend to have different travel needs deriving from the multiple tasks they must perform in their households and their communities (Grieco & Turner 1997; Hamilton 2000, in press). Low-income women also tend to be dramatically less mobile than men in the same socio-economic groups (Dutt et al. 1994).

Transport provides critical links between our homes, jobs and social lives. Mobility and travel are essential in fulfilling every role we play. Women's roles vary between societies, classes and ethnic groups. Female travel patterns vary, depending on whether women live in urban or rural areas, on the stage of economic development and on whether they are economically active.

There are, however, many common features that extend across both developed and developing societies. Even in societies where formal legal equality exists, men and women do very different work. Women are assumed to be responsible for childcare and the well-being of the household, including its health, education and housing. Managing a household includes the handling of sources of constraints (income, time) and crises (illness). Most importantly, routines and crises are coped with simultaneously. The management of travel is an integral part of the general household co-ordinating process.

Findings on women's travel patterns show that (see Figueroa 1998):

- Women make the vast majority of household trips for the purpose of shopping, taking children to school, doctors, childcare. In countries where the car is the dominant mode of transport for women's travel, such as the United States, concerns have been raised about women's potentially larger contribution to global emissions of greenhouse gases (however, see the example study below).
- Employed women make shorter work trips than men. Women contemplating employment alternatives consider the demands of household roles and other travel / employment arrangements as a strategy for family survival and as a way of securing or retaining employment.
- Car ownership in developing countries is very low. Even in car-owning households women have markedly less access to the transport resources commanded by the family.
- In developing country urban areas, the hardships of travelling by public transport in highly congested urban areas falls disproportionately on women responsible for the essential family business trips.
- Women spend more time than men on transport activities in developing country rural areas. People use human energy to transport materials in the absence of transport services. In many cultures transporting basic survival elements like water, fuelwood, agricultural tools and products is considered women's work. Again, having to walk rather than using other means of transport is keeping women out of the development process.

A recent example of studying travel patterns of different social groups is a study by Linden et al (1999): 'Gender, Traveling and Environmental Impacts'. The authors investigate travel patterns among different socio-economic groups in Sweden. According to the authors, the estimated goal for sustainable energy consumption for travelling is 11,000 MJ per person per year.

Data on gender differences in travelling patterns, i.e. distances travelled, modes of transportation and number of passengers in the car, were obtained from a database developed as part of the National Travel Survey (NTS) in Sweden. The following means of transport were examined: walking, bicycles, moped and motorcycles, car drivers, car passengers, taxi, lorry, train, commuter train, tram, bus, aeroplane and other modes of transportation. The measurement period of the NTS was from April 1994 till 1999. The study used data from the year 1996. The sample covers 50,000 people over the age range of 6-64 years.

According to the findings of their study: Elderly persons, persons with low income and women in general do not travel extensively. The ones who come closest to the goal for sustainability are elderly women in the age group of 75-84, who consume 12,000 MJ energy per annum. Middle-aged persons, persons with high income and men travel much more. Men in the age group of 45-54 years with high incomes consume the most energy, i.e. 94,000 MJ per annum. This is in contrast to men with low incomes who only consume 23,000 MJ energy per annum for their travelling.

According to the authors, women's lesser use of transport is related to their domestic responsibilities and choice of employment. Women, more often than men, rear young children and may abstain from working outside the home. When they do work outside the home, they may choose to work nearby. Women's employment is often within care and service sectors located in the centre of cities. Most public transportation is oriented in centre-periphery-directions, which makes it easier for women to take public transport.

Men with high incomes travel further to work and have more vehicles with a higher energy demand than women. Men have been shown to have different preferences from women regarding leisure activities. Women spend their leisure time in the neighbourhood, while men may travel to sports arenas at some distance from their homes. Moreover, high income earners often live in low density areas with detached houses located far away from subway stations. This could lessen their chances of using public transportation to work or doing daily shopping by using a bicycle or walking. This is in contrast to low income earners who live in high rise buildings where public transport is easily accessible.

Linden et al. point out that the differences in men and women's energy use cannot only be attributed to differences in employment rates but possibly also to differences in sectors of employment, ownership of a driving licence and car, and varied income levels. More men have driving licenses compared to women. The authors also stress the importance of income as influencing the amount of travel done both for work and for leisure.

The main conclusions of the study are that gender differences are so large that they cannot be ignored in the ongoing work for identifying and remedying unsustainable patterns of consumption. Moreover, the differences in energy consumption among socio-economic groups in Sweden are so large that they can be compared to differences in resource use commonly found between average citizens in developed and developing countries.

The importance of this research is that large disparities in consumption can exist even within a seemingly homogeneous and egalitarian society such as Sweden. Moreover the usefulness of this research is the identification of social groups (based on a disaggregation of gender, age and income) whose lifestyles are less sustainable when compared to other groups. The study has also examined why low income categories and women spend less time on travelling compared to high-income categories. These findings can have important policy implications for the middle-aged population of high income earners, who can be encouraged to change their travelling habits. However, policies must be accompanied by moves towards more sustainable infrastructure and means of transportation.

3.6. ENVIRONMENTAL DEGRADATION

Another area of inequity that is central to the debate on sustainable consumption, is the issue of environmental degradation and the implications for the poorest people.

The HDR 1998 focuses on over-consumption in developed countries and its effects on the poorest people. Environmental degradation is concentrated in the poorest regions and affects the poorest people. Deforestation is concentrated mainly in developing countries. For example, over the last two decades Latin America and the Caribbean have lost 7 million hectares of tropical forests. Most of this deforestation has taken place to meet the demand for wood and paper; nearly three quarters of it is used in developed countries (HDR 1998; World Energy Assessment 2000).

The literature on gender, environment and development has, however, shown that although over-consumption is caused by processes at the international and national level it is poor women in developing countries who bear the heaviest burden of environmental degradation (Agarwal, 1992; Dankelman & Davidson 1988, Hombergh, 1993).

Women as compared with men are more intensively engaged in household subsistence activities. Environmental degradation therefore makes their work loads heavier. As Agarwal puts it, (1992, p.138), because women are the main gatherers of fuel, fodder and water it is their working day that is lengthened with reduced access to essential items. Agarwal (1992) points out that the inability to obtain essential items due to environmental degradation has led to a number of suicides amongst young women in Uttar Pradesh, North India. Increased hardship in obtaining items for the household has caused tensions within the family.

3.7. MEASURES OF ENVIRONMENTAL PROTECTION

Thus, women bear the brunt of consumer shortages and it is their time that is impacted in trying to obtain essential consumer items. For example, in the West women have to balance shopping requirements with work and family. In developing countries women often have to walk long distances to acquire essential items. Thus, as Charkiewicz (1998, p.2) argues, due to the unequal gender division of labour, moves that are likely to be developed to promote sustainability can increase the already heavy workloads of women. The promotion of sustainable activities such as labour-intensive organic agriculture, reforestation, household recycling and segregation of waste, can put additional demands on the time of women. Similarly Eie (1995) argues that control measures such as eco-labelling place responsibilities on the individual consumers. This can increase women's workload in particular as they are the main shoppers.

3.8. CONSUMER AWARENESS AND BEHAVIOUR CONCERNING ENVIRONMENTAL PROBLEMS

There is a still patchy, but growing body of research into gender differences regarding environmental awareness and activities. There is evidence that women are more environmentally aware and engage more in protection activities such as recycling, re-use and environmentally conscious shopping (e.g., Shiva 1993; Kranendonk 1996).

An example of recent research is the study by the Sumitomo-Life Research Institute (1996): 'Consumers Awareness and Behaviour Concerning Global Environmental Problems and their Impact on Corporate Business Strategy in Japan'. The survey focused on four issues: 1. To what extent do consumers think they are responsible for environmental problems? 2. To what degree does their knowledge on environmental problems influence their ideas concerning responsibility and their behavioural pattern? 3. What actions do consumers take towards resolving environmental problems? 4. To what extent do consumers believe they are able to speed up industry action to improve the environment?

Gender was one of the many attributes that were examined along with age, income, socio-economic status and family size. Altogether, 2000 adult men and women aged 20 to 74 were interviewed in September 1995.

The following findings are relevant to the context of the present paper:

- Consumers' awareness regarding environmental problems varied according to age. Respondents in lower age groups have a 'strong sense of crisis' regarding environmental problems but are less willing to translate their awareness into action as compared to those in higher age groups. By contrast those in the 30-50 age group strongly believe they must do something about the environment, as they feel responsible for having caused environmental problems. Those over 60 years do not feel responsible for the current environmental crisis but their willingness to take action is strong.
- Among middle and older age groups it is women who are more likely to have a stronger level of awareness regarding the environment. Some 60% of 'environmental leaders', persons with strong awareness about the environment, are women, most of them in the 40-50 age group. Full-time housewives have played a central role in informing consumers about recycling and sustainable activities. Adult men, on the other hand, have hardly been known in the past to take part in environmental activities such as recycling or handling garbage. However, in recent years, more men have been involved in housework following an increase in the ratio of employed women.
- More women are involved in resource-conserving activities as compared to men. According to the survey, more women: avoid using paper towels, turn off lights in rooms that are not being used, adjust air conditioning to moderate temperatures and avoid using cars when it is possible to use public transportation. More women make environmentally conscious selections of daily goods, i.e., women refrain from buying drinks in non-returnable bottles, use products in refill containers, use recyclable materials from notebooks, use toilet paper made from recyclable materials, buy organically grown food, use natural products instead of synthetic products and refrain from buying plastic products. More women are likely to take environmentally conscious actions, i.e. buy products with the eco-friendly label, buy at stores selling environmentally friendly products, buy from manufactures making efforts to prevent pollution, and consider the energy required for transportation by buying preferentially local products.

I do not want to discuss here the possible reasons for differences found – e.g. fundamental, basically biologically determined, gender differences and/or gender roles causing different interests and needs. However, findings on gender differences vary, which might be due to varying roles. For example, CI ROAP (1997) found that men more often grew their own food and that these men were showing more environmentally conscious behaviour. There is a need for further empirical research and metaanalysis in this are. Overall, the varied findings suggest the importance of cross-cultural research in determining which gender groups around the world follow (un)sustainable consumption patterns. In terms of providing a basis for more specifically tailored approaches to promoting sustainable consumption behaviour, it is essential that we find out more about the attitudes and behaviour patterns of the different gender (and age, etc.) groups in societies.

4. Points for Discussion and Future Work

The concepts and evidence discussed above create a complex picture - and leave us with the question, what now? I would like to suggest a few points for further exploration. The way we address these issues and questions will be as important as finding solutions and implementing them.

What is the good life?

Equity is evidently the underlying issue to be addressed when looking at gender differences in consumption with a global perspective, linking the gender category with that of access to and control over resources. The challenge of empowering women clearly needs to be part of climate change mitigation, e.g. through participation of women in decision-making at all levels, change of gender roles, and addressing the issue of equity.

Wendy Annecke asked me in her review to this paper: "How do we even out, redistribute, and reduce energy consumption?" Some of the points put forward above address issues that need to be kept in mind, sources of information to be tapped into, and practical strategies to influence consumer behaviours. However, I believe there is a need for - yet another? - dual approach: On the one hand, we need to address some fundamental ethical questions, we need re-orientation of some basic values. In the history of humankind, societies and groups have gone through phases of shake-ups of value systems and of re-consideration of their value base. For example, we are currently observing this with regard to the lively discussions about biotechnology and genomics. We need to ask ourselves, what is a good life? Do we want a good life for all people on this planet? Do we want to believe that development needs to become sustainable? Many people in the wealthy nations - who account for most of the current consumption of energy, goods and services - seem to have forgotten this question, living in societies where materialistic values and individualistic life-styles are the most visible pattern. But value systems are not given, they are based on choices which people make.

Coming back to the 'dual approach', we need, on the other hand, to engage in numerous 'small' steps of changing consumer behaviour, providing and promoting sustainable consumption choices, providing information about environmental consequences, creating appropriate economic incentives, etc. These 'small' steps – which are in fact not small, either politically or economically – will complement reconsiderations of values and ethics. They need to be in place in order to change behaviour, as a subversive as well as a reinforcing factor. Given the fact that values influence actual choices only to a limited extent, new and different values will have to be reinforced by education, regulation, and the creation of new social images and desirable identities – as well as by providing information, choices and access.

Governance

How could we bring such parallel processes about? This is fundamentally a question of governance, and I would like to argue for a multi-stakeholder consultation approach. In human history, we have tried and experienced various governance approaches and systems. There are not many that we haven't tried, among them multi-stakeholder approaches, a process implying new forms of interaction, leadership, decision-making, and responsibility. A multi-stakeholder approach develops because stakeholders – i.e. all those affected by an issue and all those who are able to influence it – identify a problem which needs to be addressed, such as climate change. Given the political space and a transparent link into the government decision-making process, stakeholder groups such as industry, NGOs, civil citizens organizations, trade unions, experts, etc. can come together and consult about the reasons and possible solutions for a given problem. In the context of the issues of gender, poverty and consumption, these would be women's organizations and grassroots women's groups, climate change experts with various specializations, representatives from all parts of societies (Global North and Global South), governments, specialized agencies and local authorities, etc. They could then become part of pilot projects implementing identified strategies, monitoring progress and outcomes, and developing necessary adjustments.

All stakeholders, including governments, need to take on the responsibility of asking the questions about the good life and about what we want to believe in, what equity and justice should mean in practical terms in the lives of people. We need the information, expertise, perspectives and creativity of all stakeholders to address the fundamental questions about our value choices and identify practical strategies and solutions. Without that, we will not have the necessary expertise to adequately address the problems. Maybe even more importantly, we will also not have the necessary commitment to the implementation of necessary changes. Any radical strategies that involve sharing between rich and poor and changes of the power relations between women and men would require enormous mobilization of commitment.

An open, transparent, democratic and accountable approach to universal participation in problem-solving and implementation has not been tried often in human history. The global, complex and extremely challenging problems we are facing today, such as climate change, justify the social, political and financial investment into developing a new approach.

Social Equity vs. Environmental Protection

Looking at consumption opportunities and consumption choices of women and men, Global South and Global North points to some fundamental questions. On the one hand, poverty can lead to unsustainable or environmentally damaging behaviour, the most prominent example being the cutting down of trees for fire wood. In such cases, pursuing the goal of social equity and the eradication of poverty goes hand in hand with the goals of environmental protection: Alleviating poverty could include providing modern technologies for cooking and heating and thus limit environmental damage and greenhouse gas emissions.

On the other hand, more 'sustainable' behaviour patterns of women (and/or poor people) are due to lack of access and opportunities for (over-)consumption. In such cases, social equity and poverty eradication goals and environmental protection goals do not necessarily go hand in hand. As we know, we cannot strive to provide consumption opportunities for all citizens of the world as they are available in the highly developed countries today. Consumption levels in the North are correctly labelled 'over-consumption' and need to change. However, poverty levels are scandalous and need to change urgently to provide all human beings with fair opportunities for a decent life and life expectancy.

To achieve sustainable and just consumption levels, there is an urgent need for change both in the Global South and the Global North the resources for which need to be invested by the developed countries. Tackling over-consumption will be as difficult as tackling poverty eradication. To identify feasible strategies and to implement them will require a conserted effort of all stakeholders, including the full and equal participation of women at all levels. It will require re-thinking basic cultural values in the Global North.

Analysing Consumer Behaviour

As pointed out in the present paper, gender is an important social category determining consumer behaviour. However, it is not the only one – age, income level, ethnic groups, religious groups, etc. can be important determinants as well. Therefore, analysis of consumer behaviour needs to be conducted with a view to differences between social categories. Relevant social categories and their specific behavioural and motivational patterns need to be identified. Researchers, policy makers and stakeholders working on production and consumption patterns should integrate a strong focus on consumer behaviour and different consumer groups in their work. Social structural issues should be integrated into indicators that are being formulated to measure changes in consumption patterns. Indicators should reflect regional, national and global realities. The general aim should be to provide more information on the consumption behaviour of different social groups. It is equally important to conduct surveys that examine what factors and motivations contribute to consumer awareness, behaviour and positive actions with regard to the sustainable practices to complement the ongoing work.

Factors that influence consumption patterns, such as poverty, the gendered nature of poverty, rural-urban migration, international migration change, trade and investment, advertising, globalization and tourism, need to be further clarified. Household decision-making and the gendered division of labour are also factors which need to be taken into account when formulating strategies towards sustainable consumption. It will also be important to analyse the consequences of deconstruction of gender roles.

Influencing Consumer Behaviour

In general, consumption choices can be influenced by values and ethical considerations, such as environmental ethics, information and knowledge, the availability of affordable choices, and the social meanings and images associated with different choices.

Based on identification of relevant social categories, activities aiming to influence consumption behaviour and increase sustainable consumption can be specifically tailored towards the information base, access, motivation, status definition and stereotypical images of the specific categories and groups in question. The data, knowledge and experience of commercial advertising and marketing companies on how to effectively target consumers to strengthen their efforts in promoting sustainable consumption (for example, sensitize consumers in order to promote sustainability) should be brought into the discussion. Marketing companies have extensive and detailed knowledge about different consumer groups and have expertise with regard to influencing consumer behaviour.

Also, we need to build on practical experience: success factors of gender-specific campaigns that have aimed to promote fairer consumer rights and sustainable consumption practices should be documented and analysed, in order to enable governments and other stakeholders to use successful strategies in the future.

It should be recognized that changing the existing power relations between women and men is crucial for attaining sustainable consumption. Evidence of gender inequities in terms of access to resources, choices, and information as well as social roles and images causing (over-)consumption choices, which have been briefly outlined above, prove that gender needs to be given a major focus when aiming to influence consumption behaviour towards more sustainable choices.

Impact of Environmental Protection Measures

Policies, new technologies and measures that aim to promote sustainability must be cautioned about the already heavy work burdens that women shoulder. As women are still responsible for the majority of housework and shopping world-wide, activities such as recycling, handling of garbage and buying sustainable products can impact on their time. New technologies and measures should aim to promote equity and the emancipation of women. Services and facilities which help consumers to integrate recycling, re-use and other practices into their daily routine, should be put into place at local community level.

5. References

Agarwal, B., 1992: The Gender and Environment Debate, Lessons from India. Feminist Studies 18 (see footnote 1).

Anker, R., 1998: Gender and Jobs. Gender Segregation of Occupations in the World. Geneva, ILO.

Beckman, S., 1997: Women as Consumers. In: *The Encyclopaedia of the Consumers Movement* [Brobeck et al (eds.)]. Santa Barbara.

Bruce, J. and D. Dwyer, (eds), 1988: A Home Divided: Women and Income in the Third World. Stanford Univ. Press, Stanford.

Charkiewicz, E., 1998: Why a Gender Analysis is Important in Developing Effective Policies for Sustainable Consumption and Production. Tools for Transition, Amsterdam, Netherlands.

Consumer Unity and Trust Society India, 1996: Sustained Advertising Promoting Unsustainable Consumption. Briefing Paper, No. 2/April.

Consumers International, 1997: A Discerning Middle Class? A Preliminary Enquiry of Sustainable Consumption Trends in Selected Countries in the Asia Pacific Region. Consumers International. Regional Office for Asia and Pacific.

Costa, J. (ed.), 1994: Gender Issues and Consumer Behaviour. Sage, London.

Dankelman and Davidson, 1988: Women and Environment in the Third World: Alliance for the Future. Earthscan in association with IUCN. London.

DasGupta, P., 1993: An Inquiry into Wellbeing and Destitution. Oxford University Press, London.

DER SPIEGEL, 2000: Personalities: Maneka Gandhi / Calvin Klein / Gwyneth Paltrow. No 9, 28/02/2000, p260. Hamburg, Germany.

Dreze, J. & A. Sen, 1995: India. Economic Development and Social Opportunity. Oxford University Press, India.

Dowd, L., 1996: Sustainability, Equity and Women. In: *Initiatives for a Healthy Planet* [Kranendonk, M. (ed)]. Conference Report, NGO Forum on Women, Beijing, 1995. Published by the Wuppertal Institute, Germany.

Eie, E., 1995: Sustainable Production and Consumption. Key Note Speech for the 'Web', NGO Forum, Beijing, 1995. FOKUS - Forum for Women and Development, Oslo, Norway.

Figueroa, M., 1998: Women, Transport, Energy and the Environment. ENERGIA News, May 1998, 1-4.

Firat, A.F., 1994: Gender and Consumption: Transcending the Feminine. In: *Gender and Consumer Behaviour* [Costa, J. (ed.)]. Sage, London.

Firat, A. F., 1991: Consumption and Gender: A Common History. In: *Gender and Consumer Behaviour* [Costa, J. (ed.)]. Proceedings of a Conference on Gender and Consumer Behaviour, Salt Lake City, Utah, USA.

Friends of Earth UK, 1998: Children, Television, Advertising and the Environment. Briefing Paper By Anne Thomas. Friends of the Earth UK, London, UK.

Grazia and Furlough (eds.), 1996: The Sex of Things: Gender and Consumption in Historical Perspective. Berkeley, University of California Press.

Grieco, M. and J. Turner, 1997. Gender, Poverty and Transport: A Call for Policy Attention. In International Forum of Urban Poverty, Florence, November 1997: UNCHS (Habitat). quoted from: Barter, A.R.P. 1998. Transport and Urban Poverty in Asia: A Brief Introduction to the Key Issues. Paper presented at the UNCHS (Habitat) Regional Symposium on Urban Poverty in Asia. Fukuoka, Japan, 27 - 29 Oct 1998.

Grover, S., C. Flenley, and M. Hemmati, 1999: Gender & Sustainable Consumption. Bridging Policy Gaps in teh Context of Chapter 4, Agenda 21 'Changing Consumption and Production Patterns. UNED-UK, London.

Hamilton, K., 2000 (in press). Transport Gender Audit for the UK. London.

Haq, U.M. (ed.), 1997. Human Development in South Asia. Oxford University Press.

Hemmati, M. and CSD NGO Women's Caucus, 2000: Women & Sustainable Development 2000 - 2002. Recommendations in Agenda 21 and Suggestions for a Review of Implementation. CSD NGO Women's Caucus, London / New York.

Hombergh, H., 1993: *Gender,* Environment and Development - A Guide to the Literature. Published for INDRA, Netherlands.

Horowitz and Mohun (eds.), 1998: His & Hers. Gender, Consumption and Technology. University Press of Virginia, USA.

Kabeer, N., 1997: *Women, Wages and Intra-Household Power Relations in Urban Bangladesh*, Development and Change, 28.

Kranendonk, M. (ed.), 1996: Initiatives For a Healthy Planet. Wuppertal Institute, Germany.

Linden, A. et al, 1999: Gender Travelling and Environmental Impacts. Society and Natural Resources, 12.

Linden, A et al, 1998: Differences in Resource Consumption and Lifestyles - What are the Implications for Sustainability?, Ecological Economics Bulletin, 3(2, 2nd Quarter).

Lubar, S., 1998: Men/Women/Production/Consumption/. In: *His & Hers. Gender, Consumption and Technology*, [Horowitz and Mohun (eds.)], University Press of Virginia, USA.

MacDonald, M., 1995: Representing Women. Myths of Femininity in the Popular Media. Arnold, London.

Mackie, D.M. and D.L. Hamilton, (eds.), 1993: Affect, Cognition, and Stereotyping. Interactive Processes in Group Perception. Academic Press, San Diego, California.

Mawle, A., 1996: Why Chlorine in Tampons Matters. In: *Initiatives for a Healthy Planet* [Kranendonk, M. (ed.)], Wuppertal Institute, Germany.

Mungwashu, R.S.P., 1999: Sustainable Agriculture and Land. United Nations Association Zimbabwe / Women's Operation Green 1996 - 2000. Chinhoyi, Zimbabwe. Available at http://www.uned-uk.org/wcaucus/csdngo.htm.

Robinson and Goodman, 1996: The New Rich in Asia. Routledge, London.

Shiva, V. (ed.), 1993: Minding Our Lives. Women from the South and North Reconnect Ecology and Health. Kali for Women, New Delhi, India.

Smith, E.R. and D.M. Mackie, 2000: Social Psychology. Philadelphia, USA, Psychology Press / Taylor and Francis.

Sumitomo-Life Research Institute, 1996: Consumers Awareness and Behaviour Concerning Environmental Problems and their Impact On Corporate Business Strategy in Japan. National Institute for Environmental Studies, Japan.

Thomas, D., 1990: Intra-household Resource Allocation: An Inferential Approach. Journal of Human Resources, 25(see footnote 4).

Tsikata, D., 1995. Effects of Structural Adjustment on Women and the Poor. Third World Network, Africa Secretariat, Accra, Ghana. Available at http://www.twnside.org.sg/souths/twn/women.htm.

United Nations Department for Social and Economic Affairs, 1998: Measuring Changes in Consumption and Production Patterns. A Set of Indicators. United Nations, New York.

United Nations Development Programme, 1995: Human Development Report 1995. Oxford University Press.

United Nations Development Programme, 1998: Human Development Report 1998. Oxford University Press.

Organization for Economic Co-operation and Development, 1997: Sustainable Consumption and Production. Paris, OECD.

Vajpayi, R., 1996: Indian Women and Consumer Education. In: *Initiatives for a Healthy Planet* [Kranendonk, M. (ed.)], Wuppertal Institute, Germany.

Wells, T. and F. Sim, 1987: Till They Have Faces. Women as Consumers. International Organisation of Consumers Unions, Regional Office for Asia and Pacific Region.

World Commission on Environment and Development, 1987. Our Common Future. Oxford University Press.

World Energy Assessment Overview, 2000: New York, disseminated at the UN Commission on Sustainable Development Expert Group Meeting on Energy and Sustainable Development, March 2000.

Zanna, M.P. and J.M. Olson, (eds.), 1994: The Psychology of Prejudice. The Ontario Symposium, Vol.7. Erlbaum, Hillsdale, New Jersey.

Acknowledgement

I want to thank Shalini Grover for her work on UNED-UK's report on 'Gender & Sustainable Consumption' (Grover et al 1999). When writing the present paper, I have drawn on her research and the discussions we had in the project team with Shalini and Clare Flenley, UNED Forum. I also want to thank Markus Paul, University of the Saarland, Saarbruecken, Germany, for searching the PsycLit and Psyndex databases for relevant psychological research, and Jasmin Enayati, UNED Forum, for helping to analyse the results of the search efforts. Wendy Annecke and Raban Chanda, who have reviewed the paper, offered many useful insights and challenging questions which I am sure I have not been able to address sufficiently.

[illegible] Status and Relative Consumption

[illegible]

1. Introduction

[illegible] of environment and development to integrate the consumption [illegible] goods and services as a [illegible] contributor to human well-being [illegible] property rights and the [illegible] of environmental [illegible] allocation of natural resources ([illegible] and Oates, 1988) [illegible] applied economists seek to measure the monetary value of environmental quality in order to [illegible] costs and benefits of environmental [illegible]

The underpinnings of this approach are, of course, well understood in economic theory [illegible] preferences are [illegible] determined [illegible] that are largely independent of prevailing economic conditions. It also assumes that the costs and benefits of [illegible] individual [illegible]

[illegible] ecological economics, conceptual frameworks and empirical findings in these disciplines suggest that individuals' sense of identity, and hence the perceived value they attach to [illegible] consumption [illegible] strongly affected by aggregate patterns of economic activity ([illegible] 1987; Duesenberry, 1992). In particular, a concern for achieving a high level of relative consumption is widely viewed as an important motive behind economic behaviour (Scitovsky, 1976; Duesenberry, 1949).

The relative consumption hypothesis was introduced to economics as early as Adam Smith (1776) who observed that although the English of his day would be ashamed to appear in public without wearing shoes and linen shirts, these items were seen as unnecessary luxuries in less wealthy parts of Europe, where fashion conventions defined less costly clothing as socially acceptable. Of course, Thorstein Veblen (1899) wrote eloquently about the social waste caused by 'conspicuous consumption', in which the affluent households of his day engaged in status competition through expenditures on (conspicuous) luxury goods.

[illegible] (eds.), [illegible] Climate Change Mitigation, 191–205.

Climate Change and Relative Consumption

Richard B. Howarth

1. Introduction

Conventional models of environment and development interpret the consumption of market goods and services as a strictly positive contributor to human well-being. Such models begin with the premise that poorly defined property rights and the non-exclusive, non-rival character of environmental goods imply a role for public policies to ensure the optimal allocation of natural resources (Cropper and Oates, 1992). In microeconomic terms, applied economists seek to measure the monetary value of environmental quality in order to balance the costs and benefits of environmental degradation and thus achieve economic efficiency (Pearce, 1993). In macroeconomic terms, natural resource accountants seek to adjust measures of net national product to account for the flow of services generated by natural systems and the impacts of resource depletion on "natural capital" (Atkinson *et al.*, 1997; Costanza, 1997). Each of these practices rests on the assumption that monetary measures provide a suitable index of individuals' welfare or preference satisfaction. The problem is to express the value of market and nonmarket goods in commensurable units.

The underpinnings of this approach are of course well understood in economic theory (Just *et al.*, 1982). The model assumes that individual preferences are exogenously determined by biological and/or cultural factors that are largely independent of prevailing economic conditions. It also assumes that the costs and benefits of market goods are rival and exclusive so that an individual's consumption affects her own, and *only* her only, subjective well-being. These assumptions are useful in providing a parsimonious representation of consumer decision-making. They are, however, in strong tension with the insights of anthropological, sociological, and psychological research. The conceptual frameworks and empirical findings of these disciplines suggest that individuals' sense of identity – and hence the perceived value they attach to material consumption – is strongly affected by aggregate patterns of economic activity (Argyle, 1987; Dittmar, 1992). In particular, a concern for achieving a high level of *relative consumption* is widely viewed as an important motive behind economic behaviour (Scitovsky, 1976; Duesenberry, 1949).

The relative consumption hypothesis was introduced to economics as early as Adam Smith (1776), who observed that although the English of his day would be "ashamed to appear in public" without wearing shoes and linen shirts, these items were seen as unnecessary luxuries in less wealthy parts of Europe, where fashion conventions defined less costly clothing as socially acceptable. Of course, Thorstein Veblen (1899) wrote eloquently about the social waste caused by "conspicuous consumption," in which the affluent households of his day engaged in status competition through expenditures on perceived luxury goods.

E. Jochem et al. (eds.), Society, Behaviour, and Climate Change Mitigation, 191–206.

More recently, Richard Easterlin (1974) reviewed empirical data on the links between income and subjective well-being, as measured through surveys on self-reported happiness, in a broad sample of developing and industrial societies. Easterlin found that at any given time in any given society, those with high incomes tended to report higher levels of life satisfaction than those with lower levels of income and consumption. Conversely, however, Easterlin's data suggested that there was no correlation between a society's average income and the average well-being of its inhabitants. And in wealthy nations such as Japan and the United States, a typical person's life satisfaction remained unchanged between the 1950s and the 1970s despite quite robust economic growth.

The results surrounding this so-called "Easterlin paradox" are controversial and are subject to various interpretations (see Oswald, 1997). Ronald Inglehart (1996), for example, presents evidence of a strong positive relationship between income and welfare in societies with annual incomes below $ 6,000 per capita measured in 1991 prices. In more wealthy societies, however, Inglehart broadly confirms Easterlin's finding of a weak relationship between aggregate consumption and social well-being. In affluent societies – i.e., in those where basic needs for goods such as food, sanitation, health care, housing, and education are fulfilled – it would appear that the perceived benefits of income and consumption are defined in largely relational terms. It is perhaps more important to enjoy a high standard of living in comparison with one's peers than to achieve a particular level of absolute consumption.

Further insights on this subject are provided by the work of Solnick and Hemenway (1998), who used survey methods to examine people's preferences regarding trade-offs between absolute and relative consumption. As Frank (1985) argues, a concern for social status may cause people to choose employment and housing options in which they enjoy high relative status in comparison with their co-workers and neighbors, even if this choice is linked to some reduction in aggregate income. To test this hypothesis, Solnick and Hemenway asked a sample of human subjects to compare a scenario in which they would earn a real income $ 50,000 per year in a society with an average income of $ 25,000 with an alternative in which they would earn $ 100,000 per year in a society with an average income of $ 200,000. Roughly half of the respondents favoured each option, suggesting that individuals may be willing to give up quite substantial amounts of income if doing so enhances their relative standing.

The implications of status-seeking behaviour for environmental management are explored by Howarth (1996), who considers the interplay between relative consumption effects and an environmental disamenity in a simple static model (see also Ng and Wang, 1993; Brekke and Howarth, 2000). Although conventional theory asserts that consumption taxes distort the efficiency of resource allocation to the detriment of human well-being, Howarth finds that a concern for relative consumption implies that standard market goods impose negative externalities that can be internalized through a consumption tax. In a similar vein, Frank (1999) advocates a graduated consumption tax with a top rate of 90% to reduce the social inefficiencies of status seeking. Frank's recommendation, however, is heuristic in nature and is not grounded on a formal empirical analysis.

Although conventional theory suggests that environmental externalities should be taxed according to individuals' marginal willingness to pay for environmental conservation, Howarth finds that standard measures of willingness-to-pay understate the social benefits of environmental amenities in the presence of relative consumption effects. In this context, individuals attach excess weight to the value of their personal consumption, inducing a bias towards excess consumption and environmental degradation. Restoring the efficiency of resource allocation requires careful attention to the interplay between these coupled externalities, not the mere measurement of environmental values in monetary terms.

The purpose of this chapter is to investigate the implications of these issues in a simplified dynamic model of the links between climate change and the world economy – specifically, Nordhaus' (1994) Dynamic Integrated model of Climate and the Economy (DICE). The analysis is designed to assess environmentalists' claim that the pursuit of social status in a market economy provides incentives to pursue inefficiently high rates of economic growth to the detriment of community life and environmental quality – public goods that require strong institutions for collective provisioning (Durning, 1992). To accomplish this task, the chapter incorporates the conceptual framework of Howarth's static model in an explicitly dynamic specification that is numerically calibrated to accord with the stylized facts surrounding tradeoffs between economic growth and climate stabilization. Although the analysis rests on a number of assumptions that might be subjected to further examination, it shows that the environmentalist perspective may be reconciled with a formal model of economy/environment interactions under plausible assumptions regarding the empirical importance of relative consumption effects.

2. The Dice Model: Base Specification

The DICE model offers a parsimonious representation of the links between economic growth, greenhouse gas emissions, and global climate change. In the standard version of this model, resource allocation decisions are chosen to maximize the objective function:

$$(1) \quad \sum_{t=0}^{\infty} N_t \ln(c_t) / (1+\rho)^t$$

subject to a set of technological and environmental constraints. In this formulation, N_t is total population at a sequence of dates $t = 0,1,2,\ldots$; c_t is the consumption of a typical individual; and ρ is the pure rate of time preference. Nordhaus uses this approach to simulate the aggregation of individual and collective choices in the world economy.

The production possibilities of the model may be summarized using the reduced-form equation:

(2) $$N_t c_t + K_{t+1} = f(K_t, L_t, E_t, T_t, t) + K_t .$$

In this formulation, $N_t c_t$ is aggregate consumption, K_t is the capital stock, L_t is the supply of labor, E_t is the emission of greenhouse gases, and T_t is the increase in mean global temperature relative to the pre-industrial norm. The net production function $f(\cdot)$ is concave with respect to capital, labor, greenhouse gas emissions, and temperature; linearly homogeneous and increasing in capital, labor, and emissions; and decreasing in temperature. Time, which is measured in decades, enters the production function to allow for exogenous technological change.

The supply of labor is given by:

(3) $$L_t = \lambda_t N_t$$

where λ_t represents labor effort, or the proportion of people's time devoted to work activities. In Nordhaus' specification, λ_t is taken as fixed so that labor services are strictly proportional to population.

Finally, temperature change is determined by past greenhouse gas emissions so that:

(4) $$T_t = T(E_0, \ldots, E_{t-1}) .$$

Current emissions give rise to increases in future temperatures so that $\partial T_t / \partial E_{t-i}$ is positive for all $i > 0$. The function $T(\cdot)$ is a concave in all of its arguments.

The parameterization of this model is fully described by Nordhaus (1994) and need only be sketched here. In DICE, population grows from 5.6 billion persons in 1995 to 10.5 billion in the long-run future. The bulk of this increase is concentrated in the next one hundred years. The pure rate of time preference (ρ) is set equal to 3% per year. The production function exhibits the familiar Cobb-Douglas form defined over inputs of capital and labor, adjusted to reflect the costs and benefits of greenhouse gas emissions. The capital stock depreciates at an annual rate of 10%, and the parameters of the production function are calibrated to account for the fact that labor accounts for three-quarters of the value of output while capital accounts for the remainder. Technological change augments total factor productivity at an initial rate of 1.0% per year, but slows gradually over time to reflect observed long-term trends. These assumptions ensure that the model replicates observed rates of interest, investment, and economic growth in baseline simulations.

The labor supply parameter λ_t is defined only implicitly in DICE, though the value of this parameter is important in the elaboration of the model that is described below. According to the United Nations Development Programme (1993), a typical worker labors some 38 hours per week in nations with detailed labor statistics. On the assumption that individuals are active for 16 hours each day, or 112 hours each week, it follows that the proportion of time spent working is $\lambda_t = 0.34$.

DICE assumes that a doubling of atmospheric greenhouse gas concentrations relative to the pre-industrial norm would lead to a 2.9°C temperature increase and a 1.2% loss of gross world product. Climate change damages increase with the square of the temperature change caused by human activities. The model is premised on the assumption that greenhouse gas emissions would rise from 9.3 to 24 billion tonnes of carbon equivalent (tce) over the next century in the absence of emissions control measures. A 50% emissions reduction leads to a 0.9% loss of gross world output, while control costs rise to 6.9% with the complete control of emissions. Once in the atmosphere, greenhouse gas molecules have a residence time of 120 years, with a doubling of concentrations occurring by the mid-twenty-first century in the absence of abatement policies.

The main findings of the DICE model are well-known in the literature. As illustrated in Table 1, the optimal path for this economy supports substantial growth in aggregate consumption, the capital stock, and greenhouse gas (GHG) emissions. Under the stipulated assumptions, it is better to bear the costs of climate change than to impose aggressive greenhouse gas emissions abatement policies. Hence emissions are reduced by only 9-14% in comparison with unconstrained levels, while mean global temperature increases by 4.6°C through the year 2195.

Table 1: DICE model base run

		1995	**2035**	**2075**	**2115**	**2155**	**2195**
Population	10^9 persons	5.5	7.9	9.2	9.9	10.3	10.4
Consumption	10^{12} 1989 USD/yr	19	45	73	96	112	123
Capital stock	10^{12} 1989 USD/yr	48	120	202	273	325	361
Labor effort	%	34	34	34	34	34	34
GHG emissions	10^9 tce/yr	8.5	14	19	22	24	25
Temperature change	°C	0.8	1.6	2.5	3.4	4.1	4.6

3. Relative Consumption Effects

The discussion thus far has been limited to a presentation of the core structure and results of DICE under Nordhaus' base specification. We turn now to the description of a more general version of this model that incorporates relative consumption effects and a labor-leisure tradeoff. In the specification of preferences, suppose that the welfare of a typical household is captured by the objective function:

$$W = \sum_{t=0}^{\infty} N_t u_t / (1+\rho)^t \tag{5}$$

where:

$$(6) \quad u_t = (1-\alpha)\ln(c_t) + \alpha\ln(s_t) + \beta\ln(1-\lambda_t).$$

In this setting, u_t is the utility of a typical individual, while $s_t = c_t / C_t$ is an index of her relative consumption in comparison with the social average C_t. Under these assumptions, the parameter α defines the weight individuals attach to their perceived economic standing. The third term of the utility function reflects the disutility of labor or, equivalently, the utility gained from leisure. Here $1-\lambda_t$ represents the proportion of time that a typical person devotes to non-work activities, which is explicitly chosen to balance the perceived costs and benefits of labor.

A first question about this model concerns its ability to replicate the core findings of DICE in the absence of relative consumption effects. To shed light on this question, we set $\alpha = 0$ and consider the optimal resource allocation that arises when the objective function W is maximized subject to the model's technological and environmental constraints. The solution to this problem of course depends on the value that is selected for the preference parameter β, which defines the relative weight a typical person attaches to the enjoyment of leisure. Sensitivity analysis establishes that a value of $\beta = 1.8$ gives rise to an optimal resource allocation in which the numerical results of the current specification conform quite closely to those generated by Nordhaus' base model (Table 2). Accordingly, this parameter value is employed throughout the ensuing analysis.

Table 2: $\alpha = 0$

		1995	**2035**	**2075**	**2115**	**2155**	**2195**
Population	10^9 persons	5.6	7.9	9.2	9.9	10.3	10.4
Consumption	10^{12} 1989 USD/yr	19	45	72	94	110	120
Capital stock	10^{12} 1989 USD/yr	48	120	199	267	318	352
Labor effort	%	34	34	33	33	33	33
GHG emissions	10^9 tce/yr	8.5	14	19	22	24	25
Temperature change	°C	0.8	1.6	2.5	3.4	4.1	4.6

The correspondence between the results given in Tables 1 and 2 suggests that the assumption of a fixed relationship between labor supply and population is a useful descriptive device under the stipulated conditions. It further suggests that DICE may be understood as a special case of the more general model described in this chapter. Against the background, we are ready to consider the implications of relative consumption effects for the efficiency of resource allocation in a competitive economy.

3.1. COMPETITIVE EQUILIBRIUM

The most common approach to modelling a market economy is to disentangle the problems of utility and profit maximization so that the role of prices in shaping resource allocation decisions is brought sharply into focus. For our purposes, however, it is useful to take a more streamlined approach in which we assume that production decisions are managed directly by a representative household. This approach is reasonable in at least two respects. First, firms in actual economies are (in theory) managed to maximize shareholder value, which captures the contribution that a firm's profits make to its owners' well-being. Small businesses, of course, are managed by their proprietors in precisely this manner. Second, the results obtained using this approach correspond exactly to those that arise from the more typical formulation in which utility and profit maximization are carried out independently by households and firms.

We suppose that the budget constraint faced by a typical household may be written in the form:

$$(7) \qquad (1+\tau_t)N_t c_t + K_{t+1} = f(K_t, L_t, E_t, T_t, t) + K_t + \pi_t - v_t E_t .$$

In this specification, τ_t is a unit tax on consumption that is levelled by an exogenous agency – the government – to internalize the externalities generated by status-seeking; v_t is a Pigouvian tax on greenhouse gas emissions; and π_t is a lump-sum transfer that the government uses to release the revenues generated by the consumption and emissions taxes. This constraint implies that the value of net output plus transfer income is divided between expenditures on consumption, investment, and tax payments.

In the context of a competitive economy, it is natural to assume that individuals take tax rates, income transfers, the average consumption level in society, and environmental conditions as fixed and beyond the scope of their personal decisions. Under this assumption, the maximization of household welfare (*W*) subject to equation (7) gives rise to the first-order conditions:

$$(8) \qquad \frac{\beta c_t}{1-\lambda_t} = \frac{\partial f_t / \partial L_t}{1+\tau_t}$$

$$(9) \qquad \frac{(1+\rho)c_{t+1}}{c_t} = \left(1+\frac{\partial f_{t+1}}{\partial K_{t+1}}\right)\left(\frac{1+\tau_t}{1+\tau_{t+1}}\right)$$

$$(10) \qquad v_t = \frac{\partial f_t}{\partial E_t} .$$

In the first equation, the marginal rate of substitution between consumption and leisure is equated with the after-tax marginal productivity of labor, or the increase in consumption that a marginal decrease in leisure provides from the household's point of view. The second condition equates the marginal rate of intertemporal substitution with the gross return to capital investment, which again is adjusted to account for the impacts of taxation. Finally, economic actors choose the level of greenhouse gas emissions to equate the marginal productivity of emissions with the prevailing emissions tax.

Given a set of feasible policies concerning consumption taxes, emissions taxes, and lump-sum transfers, the conditions described above give rise to a well-defined competitive equilibrium. In equilibrium, the average consumption level is equated with the consumption level enjoyed in a typical individual so that $c_t = c_t$. In addition, it is useful to restrict attention to the case where the government's budget is balanced in each period so that $\pi_t = \tau_t N_t c_t + v_t E_t$. This assumption eliminates the need to consider the complications that arise when the government incurs a deficit or surplus that must be financed through capital markets.

3.2. OPTIMAL TAXATION

As was noted above, the model as specified in this chapter involves a coupling between two types of externalities. First, current greenhouse gas emissions impose costs on the future economy through the damages caused by increased mean global temperature. Second, each individual's consumption raises the average consumption level in society, thereby diminishing the perceived social status and subjective welfare of all other persons. In the face of these externalities, the taxation of emissions and consumption is justified to improve the efficiency of resource allocation. We thus turn our attention to the identification of optimal tax policies.

Consider the problem faced by a social planner who sought to maximize the perceived welfare of a typical household (*W*) subject to the model's technical constraints as embodied in equations (2)-(4). Recognizing that $c_t = c_t$ so that a typical person's social status is independent of the prevailing consumption level (i.e., $s_t = c_t / \bar{c}_t = 1$), this problem yields the optimality conditions:

$$\frac{\beta c_t}{(1-\alpha)(1-\lambda_t)} = \frac{\partial f_t}{\partial L_t} \tag{11}$$

$$\frac{(1+\rho)c_{t+1}}{c_t} = \left(1 + \frac{\partial f_{t+1}}{\partial K_{t+1}}\right) \tag{12}$$

$$\frac{\partial f_t}{\partial E_t} = -\sum_{i=1}^{\infty}\left(\frac{\partial f_{t+i}}{\partial T_{t+i}}\frac{\partial T_{t+i}}{\partial E_t}\prod_{j=1}^{i}\left(1+\frac{\partial f_{t+j}}{\partial K_{t+j}}\right)^{-1}\right). \tag{13}$$

The first condition equates the marginal rate of substitution between consumption and leisure – as viewed from a social perspective that takes the average level of consumption as endogenous – with the marginal productivity of labor. A comparison of equations (8) and (10) shows that a competitive equilibrium will be Pareto efficient only if the consumption tax is chosen according to the optimal tax formula:

$$(14) \qquad \tau_t^* = \frac{\alpha}{1-\alpha}.$$

In this formulation, consumption is taxed to account for individuals' private incentive to overconsume in the pursuit of enhanced social status. The optimal tax rate rises from zero to infinity as the weight attached to relative consumption (α) increases from zero to unity.

Under the second condition for optimal resource allocation (equation (12)), the marginal rate of intertemporal substitution is set equal to the gross return on capital investment. Given the optimal consumption tax, it is easily shown that this condition is equivalent to the formula contained in equation (9) that governs competitive economies. Under these conditions, the private decisions of individual households will give rise to an optimal pattern of consumption and investment.

Finally, condition (13) equates the marginal productivity of greenhouse gas emissions with the discounted marginal costs that current emissions impose on the future economy. In this expression, the term $-\partial f_{t+i} / \partial T_{t+i}$ represents the marginal damages in period $t+i$ caused by increased mean global temperature, while $\partial T_{t+i} / \partial E_t$ represents the marginal increase in period $t+i$ temperature caused by period t emissions. The final term in this equation represents the discount factor applied to the future damages caused by climate change. Since the marginal productivity of capital $\partial f_t / \partial K_t$ reflects the rate of return or implicit discount rate that households earn on capital investments, the results hold that future costs and benefits should be discounted at the prevailing interest rate. Substitution of equation (13) into equation (10) yields the efficient emissions tax rate:

$$(15) \qquad v_t^* = -\sum_{i=1}^{\infty}\left(\frac{\partial f_{t+i}}{\partial T_{t+i}}\frac{\partial T_{t+i}}{\partial E_t}\prod_{j=1}^{i}\left(1+\frac{\partial f_{t+j}}{\partial K_{t+j}}\right)^{-1}\right).$$

This formula may be interpreted as a standard Pigouvian tax that equates the marginal benefits and discounted marginal costs of greenhouse gas emissions.

It is worth noting that this formula, which conforms to the standard prescriptions of environmental economics, holds in the case of the present model because the costs and benefits of greenhouse gas emissions are realized through their impacts on production and, therefore, consumption. If the model were respecified so that climate change imposed environmental disamenities that entered directly into people's utility functions, then equation (15) would understate the optimal emissions tax. More specifically, Howarth (1996) examines a static model that combines relative consumption effects with a pollutant that diminishes the amenities provided by environmental systems. In this setting, the optimal pollution tax obeys the formula $v^* = (1+t^*)MC$ where t^* is the optimal consumption tax and MC is the marginal cost imposed by the pollutant, measured in terms of people's willingness to pay for pollution control. In the presence of relative consumption effects, individuals overvalue consumption and undervalue the social benefits of environmental quality.

3.3. NUMERICAL SIMULATIONS

The preceding section shows that taxes on both personal consumption and greenhouse gas emissions are necessary to achieve economic efficiency when a concern for social status, defined in terms of relative consumption, is introduced to a stylized model of climate-economy interactions. This finding supports the environmentalist argument that overconsumption leads to social and ecological costs that are overlooked by conventional approaches to pollution control policy (Durning, 1992). As was noted in the introduction, there is substantial evidence that relative consumption effects are a significant component of real-world preferences. To demonstrate the relevance of this argument for climate change policy, however, requires more than existence proofs and optimal tax formulae. One must also show that the introduction of status effects substantially alters the quantitative outcomes of standard numerical models. It is to this task that we now turn.

Figures 1-4 depict the main numerical results of this analysis. These figures show the optimal time paths for consumption, labor effort, greenhouse gas emissions, and mean global temperature for various values of α, the parameter that defines the weight individuals attach to relative as opposed to absolute consumption in gauging their utility. Of course, the optimal path for $\alpha = 0$ corresponds to the case in which a standard Pigouvian tax is imposed on greenhouse gas emissions while status externalities remain unregulated. Also shown is a *laissez faire* path in which $\tau_t = v_t = 0$ so that no steps are taken to reduce status competition or greenhouse gas emissions.

As the figures show, relatively small differences exist between the *laissez faire* scenario and the optimal path that emerges in the absence of status effects. Under *laissez faire*, per capita consumption rises from \$ 3,469 in 1995 to \$ 11,454 in 2195, while greenhouse gas emissions rise from 9.3 to 28.9 billion tce. Over this period, mean global temperature rises by 4.9 °C relative to the pre-industrial norm. In comparison with this baseline, the optimal path for $\alpha = 0$ has virtually no impacts on the evolution of consumption and leisure levels and, by extension, economic welfare. The rate of greenhouse gas emissions control rises from 9% to 15% over the period of analysis, while the increase in mean global temperature is limited to 4.6 °C through the year 2195. These results are driven by the assumptions that emissions abatement would be relatively costly and that temperature changes of this magnitude would have modest economic impacts.

Figure 1: Consumption (Trillion 1989 $/year)

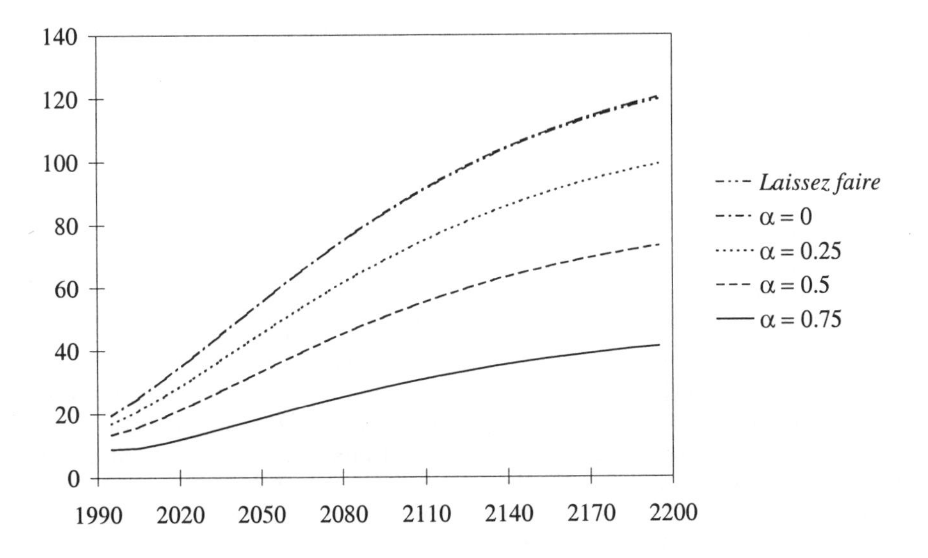

Figure 2: Labor Effort (%)

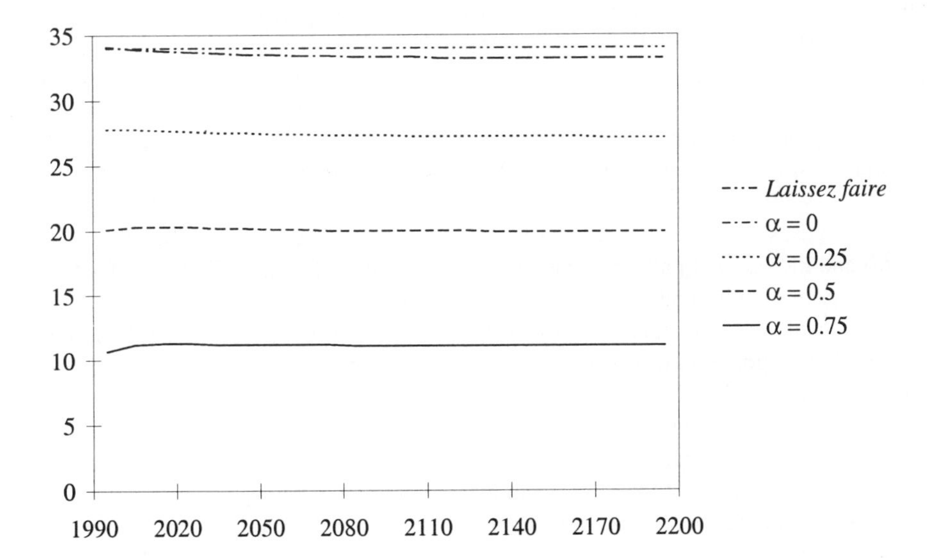

Figure 3: Greenhouse Gas Emissions (Billion tce/year)

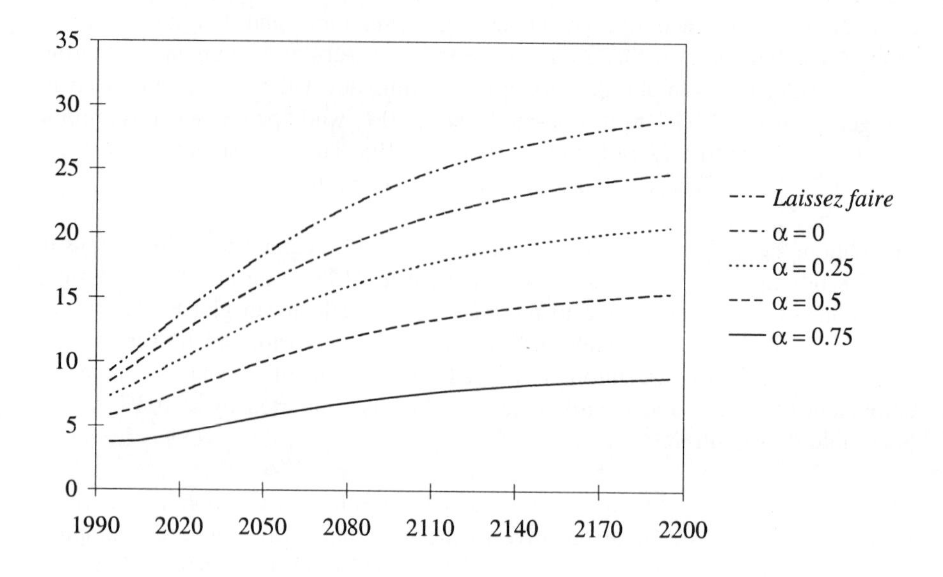

Figure 4: Temperature Change (EC)

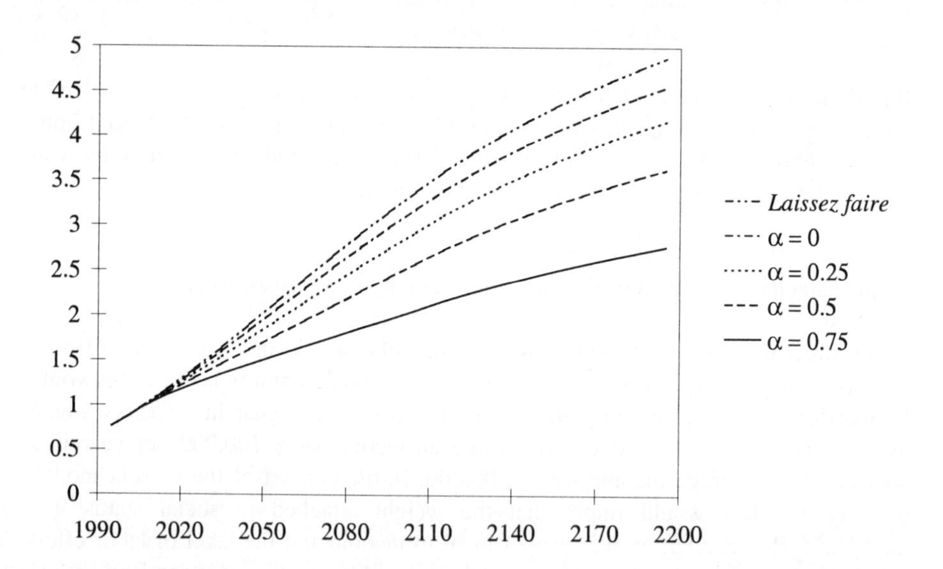

The introduction of status effects leads to important changes in these results. As one would expect, higher values of α give rise to optimal paths with decreased levels of consumption, labor activity, greenhouse gas emissions, and long-term climate change. The figures show that each of these effects is substantial in numerical terms. For $\alpha = 0.5$, for example, the proportion of time devoted to labor falls from its unregulated level of 34% to its optimal level of 20%, while per capita consumption rises from $ 2,398 to $ 7,018 between 1995 and 2195. These consumption levels are some 31-39% below those that prevail in under *laissez faire*.

In addition, the optimal path for $\alpha = 0.5$ gives rise to greenhouse gas emissions of just 5.8 billion tce in 1995 and 15.4 billion tce in 2195. With effective emissions reductions of 37 - 47% relative to the *laissez faire* path, mean global temperature increases by only 3.6 °C through the year 2195. This scenario, which promotes enhanced economic welfare through reduced status competition and the increased enjoyment of leisure and environmental quality, is supported by a 100% tax on household consumption.

This analysis suggests that optimal rates of labor effort, consumption, and environmental degradation are quite sensitive to the importance of relative consumption effects under the assumptions of this model. In the presence of status effects, imposing a Pigouvian tax on pollution is insufficient to achieve efficient levels of greenhouse gas emissions. Since status-seeking behaviour causes people to pursue excess consumption at the expense of leisure and the environment, environmental policies must be coordinated with a consumption tax to bring private decisions into alignment with the collective welfare of society.

Based on the available evidence, are there reasons for selecting a particular value of the parameter α that would identify one of these scenarios as a true "social optimum"? On the one hand, Frank (1999) advocates a graduated consumption tax with a maximum rate of 90% to internalize status externalities. In the context of the present model, a 90% consumption tax would correspond to a social optimum in the case where $\alpha = 0.47$. As was noted above, however, Frank's analysis is mainly heuristic in character, so his proposal requires further empirical justification.

More direct evidence is provided by the work of Solnick and Hemenway (1998), who as we have seen used survey techniques to establish that a typical person would be indifferent between earning an income of $ 50,000 per year in a society where average income was $ 25,000 and earning an income of $ 100,000 per year in a society where average income was $ 200,000. In the context of the present model, this stylized fact would imply that the weight attached to social status was $\alpha = 0.33$ if consumption was proportional to income and the level of labor effort was held constant under the preferences described by equation (6) (see Brekke and Howarth, 2000). Under this assumption, a 50% tax on household consumption would be necessary to internalize the externalities caused by status-seeking behaviour.

Of course, this analysis rests on particular assumptions concerning preferences that might be subjected to further investigation. The logarithmic utility function, for example, assumes a unitary elasticity of substitution between consumption, leisure, and social status. Alterations in this assumption might affect the precise results of the numerical analysis. And in structural terms, the analysis abstracts away from the existing system of taxation and public expenditures. In the United States, a marginal income tax rate of roughly 35% is faced by many middle class taxpayers. Since the impacts of this tax are roughly equivalent to those of a 54% consumption tax, one might conclude that status externalities are effectively controlled by the prevailing tax system.

It should be borne in mind, however, that economists and public officials typically view taxation as a distortionary influence that is necessary to obtain the social benefits of state-financed public goods. If a consumption tax were reinterpreted as a measure that yielded direct *benefits* by correcting the distortions associated with status seeking, then tax rates higher than those employed under existing policies would presumably emerge as socially efficient. The implications of status effects for the general problem of optimal taxation is an area that is ripe for further theoretical and empirical research.

4. Summary and Conclusions

The analysis presented in this chapter introduces a concern for social status, measured in terms of relative consumption, into a standard model of climate-economy interactions. As we have seen, there are sound conceptual and empirical reasons to believe that social status is an important object of real-world human preferences. If one accepts this premise, then the standard argument that the taxation of income or consumption introduces distortions that necessarily impair the efficiency of resource allocation is no longer conceptually valid. Instead, a Pigouvian tax on consumption may be viewed as a beneficial measure that internalizes the social costs imposed by status seeking. In the absence of such a tax, individuals have incentives to pursue excess levels of consumption at the expense of leisure and environmental quality.

To shed further light on this conclusion, the chapter describes the implications of status effects for the quantitative performance of a standard numerical model that is employed in climate change policy analysis – Nordhaus's (1994) Dynamic Integrated model of Climate and the Economy (DICE). By incorporating relative consumption effects and a labor-leisure tradeoff, the analysis finds that a 50% consumption tax might be justified based on a preliminary review of the available empirical evidence. A failure to incorporate relative consumption effects may give rise to levels of labor, consumption, and greenhouse gas emissions that substantially exceed those that would prevail given efficient resource allocation.

Of course, the analysis rests on particular assumptions regarding the characteristics of technology and preferences, and it abstracts away from the costs and benefits of the existing system of taxes and public expenditures, instead focusing narrowly on the interplay between the externalities associated with status seeking and environmental degradation. There is therefore a need for further theoretical and empirical research on the implications of status effects for economic welfare. The analysis, however, shows that the standard assumption that individual preferences are exogenously determined and unaffected by a concern for relative economic standing can substantially bias the results of applied economic modelling. Revised conceptual models are needed that reflect the more subtle dimensions of human psychology.

5. References

Argyle, M., 1987: *The Psychology of Happiness.* Methuen, New York.

Atkinson, G., R. Dubourg, K. Hamilton, M. Munasinghe, D. Pearce, and C. Young, 1997: *Measuring Sustainable Development – Macroeconomics and the Environment.* Edward Elgar, Cheltenham.

Brekke, K.A. and R.B. Howarth, 2000: The social contingency of wants – implications for growth and the environment. *Land Economics* (in press).

Costanza, R., R. d'Arge, R. deGroot, S. Farber, M. Grasso, B. Hannon, K. Limburg, S. Naeem, R.V. O'Neill, J. Paruelo, R.G. Raskin, P. Sutton, and M. van den Belt, 1997: The value of the world's ecosystem services and natural capital. *Nature*, 387, pp. 253-260.

Cropper, M.L. and W.E. Oates, 1992: Environmental economics – a survey. *Journal of Economic Literature*, 30, pp. 675-740.

Dittmar, H., 1992: *The Social Psychology of Material Possessions.* St. Martins, New York.

Duesenberry, J.S., 1949: *Income, Saving and the Theory of Consumer Behaviour.* Harvard University Press, Cambridge, Massachusetts.

Durning, A., 1992: *How Much Is Enough?* Norton, New York.

Easterlin, R.A., 1974: Does economic growth improve the human lot? In: *Nations and Households in Economic Growth* [David, P.A. and M.W. Reder (eds.)]. Academic Press, New York.

Frank, R.H., 1985: *Choosing the Right Pond.* Oxford University Press, New York.

Frank, R.H., 1999: *Luxury Fever – Why Money Fails to Satisfy in an Era of Excess.* Free Press, New York.

Howarth, R.B., 1996: Status effects and environmental externalities. *Ecological Economics*, 16, pp. 25-34.

Inglehart, R., 1996: The diminishing utility of economic growth – from maximizing security toward maximizing subjective well-being. *Critical Review*, 10, pp. 509-531.

Just, R.E., D.L. Hueth, and A. Schmitz, 1982: *Applied Welfare Economics and Public Policy.* Prentice-Hall, Englewood Cliffs, New Jersey.

Ng, Y.K. and J. Wang, 1993: Relative income, aspiration, environmental quality, individual and political myopia – why may the rat-race for material growth be welfare reducing? *Mathematical Social Sciences*, 26, pp. 3-23.

Nordhaus, W.D., 1994: *Managing the Global Commons.* MIT Press, Cambridge, Massachusetts.

Oswald, A.J., 1997: Happiness and economic performance. *Economic Journal*, 107, pp. 1815-1831.

Pearce, D.W., 1993: *Economic Values and the Natural World.* MIT Press, Cambridge, Massachusetts.

Scitovsky, T., 1976: *The Joyless Economy.* Oxford University Press, New York.

Smith, A., 1776: *An Inquiry into the Nature and Causes of the Wealth of Nations.* Strahan and Cadell, London.

Solnick, S.J. and D. Hemenway, 1998: Is more always better – a survey on positional concerns. *Journal of Economic Behaviour and Organization*, 37, pp. 373-383.

United Nations Development Programme, 1993: *Human Development Report.* Oxford University Press, New York.

Veblen, T., 1899: *The Theory of the Leisure Class*. MacMillan, New York.

Acknowledgement

This chapter is based on work sponsored by the Norwegian Research Council. I thank Kjell Arne Brekke, Phillipe Crabbé, and Ronald Wendner for thoughtful comments on a previous draft

Social interactions and conditions for change in energy-related decision-making in SMCs – an empirical socio-economic analysis

Stephan Ramesohl

Abstract

The paper discusses the contribution of socio-economic research to energy policy analysis, with emphasis on the issue of enhancing the diffusion of energy efficiency measures in industry. It presents a dynamic approach to describing the adoption process in small and medium sized firms. The implications for traditional engineering economic analyses are discussed, especially with regard to transaction costs. New priorities for policy research in the field of technical change and market transformation are derived.

1. Engineering-Economic Analyses – Do They Miss the Problem?

At the threshold to the next millenium, the global community faces serious environmental, social and economic challenges. For the industrialized countries, dramatic improvements in energy and resource efficiency, changes of industrial structures and a move towards new, less resource demanding life style patterns will be needed to enter a path of sustainable development (Sachs *et al.,* 1998). In this context, "increased efficiency in the end use of energy offers the most immediate, largest and most cost-effective opportunity to reduce consumption and environmental degradation" (Houston Declaration of the World Energy Council 1998). Beside the long-term integration of renewable energies, therefore, strategies to enhance energy efficiency will continue to represent a key element of energy and climate policy for the decades to come (European Commission, 1998). Increasing emphasis will have to be put on measures to stimulate and accelerate the market penetration and broader diffusion of energy efficiency solutions, i.e. energy efficient technologies and organizational practices such as energy management.

During the past two decades, energy policy has been strongly influenced by engineering-economic analyses, which made an important contribution to identifying technically feasible and economically efficient opportunities to save energy and to reduce GHG emissions. Bottom-up analyses have pointed to the fact that a large share of the unexploited efficiency potentials can be realized by existing technologies which are technically proven and already available on the markets.

With regard to the challenge of enhancing diffusion, however, energy policy analysis needs not only to analyse the size and structure of energy efficiency potentials, but also to examine the prerequisites and driving forces for a rapid adoption of energy saving measures by the various end-user groups. Answers have to be found to these questions:

E. Jochem et al. (eds.), Society, Behaviour, and Climate Change Mitigation, 207–226.

1. What parameters determine end-users' adoption and implementation of energy efficient technologies and organizational practices, and how can these factors be addressed by policy instruments?
2. What are the impacts and the costs of such energy policy interventions, and how can the efficiency of policy programmes be increased?

1.1. THE MICRO-ECONOMIC UNDERSTANDING OF DECISION-MAKING

In this context, the assessment of technical potential needs to be enlarged by an investigation of behavioural aspects of the adoption of new energy efficiency measures. In engineering-economic analyses, behaviour is commonly represented by a micro-economic notion of decision-making in the tradition of neoclassic economic theory. From this perspective, energy related behaviour – regardless of whether private households or industrial organizations are involved – is perceived as an economic optimization (mostly a cost minimization) undertaken by the economic agent (Fig. 1). This optimization takes place in a market environment, and adoption is described as a function of exogenous input variables such as energy prices, technology costs, legal constraints etc. Usually, engineering economic analyses build on this understanding by modelling decision as a rational, cost minimizing choice from a set of energy efficiency technologies which are ranked in the order of their marginal costs of conserved energy and carbon (conservation supply curves CSC) (Verdonck *et al.*, 1998).

Figure 1: The neoclassical micro-economic representation of the adoption process

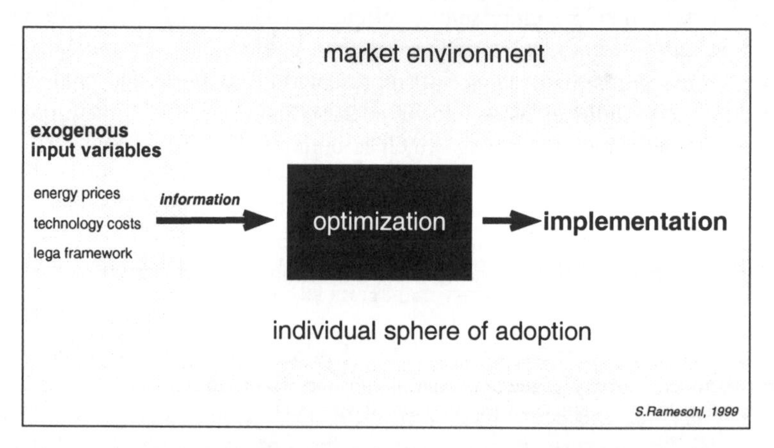

For years, this micro-economic foundation has served as the conceptual framework for energy research and policy analysis, and it has provided the forum for an engaged debate on market barriers and imperfections. Two controversial lines of argumentation can be identified: one side argues that the incomplete exploitation of apparently cost effective opportunities to save energy can be explained to a large extent by hidden costs and risk factors which are not sufficiently considered in engineering-economic analyses[1]. According to this view, apart from certain limitations mainly relating to public goods, markets are able to reach equilibrium and there is little need for policy action[2]. Others argue that the existence of untapped potentials is explained by barriers and market imperfections such as distorted energy prices, incomplete information, legal disincentives etc. which justify energy policy intervention in order to achieve socially desirable and macro-economically optimal degrees of energy efficiency[3].

1.2. SHORTCOMINGS OF THE BARRIER DEBATE

No doubt the barrier debate has been of great importance in directing scientific and political attention towards crucial problems of energy market distortions, prevailing dis-equilibria and opportunities to enhance efficiency through suitable policy instruments and programmes. At the same time, however, the discussion has been inherently restricted by the common micro-economic foundation, which provided the unquestioned conceptual background for engineering-economic analyses and the relevant policy evaluations. When looking at industry, economic decisions appear to be carried out by an impersonal "black box", because organizational, social and communicative aspects of the adoption process in firms are systematically neglected. As a consequence, decisive parameters and policy variables are excluded by definition from the analysis, so that the nature of the adoption process is not investigated but postulated *ex-ante* through the assumption of rational optimizing behaviour by "the" firm[4]. But crucial questions still remain unanswered: What is "the" firm? Who are the actors forming the social system "enterprise", and how do they interact? How can they be addressed by policy?

The traditional engineering-economic representation of behaviour and decision-making in organizations gives no answer to these questions, and is therefore of limited use for energy policy-making which is committed to enhancing diffusion. A surprisingly poor performance can be observed with regard to the provision of knowledge on the internal parameters and conditions for decision-making and behavioural changes in organizations, although these aspects are of crucial importance in understanding diffusion and market transformation in the industrial sector[5].

---

1 This discrepancy between engineering-economic analyses and micro-economic theory is commonly discussed as the energy efficiency gap, cf. Jaffe, Stavins 1994.

2 cf. Sutherland, 1991, 1994, 1996; Hasset, Metcalf 1993, 1996; Metcalf 1994.

3 cf. Fisher, Rothkopf 1989; Carlsmith et al. 1990; Jochem, Gruber 1990; Howarth, Andersson 1993; Ayres 1994; Komey, Sanstad 1994, Sanstad, Howarth 1994; Weber 1997.

4 Since the 60's, Cyert and March and others have been pointing out the flaws and limitations of the neoclassical approach to organizational behaviour, cf. Cyert, March 1992; March, Simon, 1993, March 1994.

5 For some of the rare exceptions see DeCanio 1993, 1998; Gillissen 1994; Gillissen, Opschoor 1994.

In my understanding, there is an urgent need to overcome the methodology-immanent limitations of policy evaluation and the underlying engineering-economic analyses, in order to derive a deeper understanding of the adoption of technical and social innovations by firms, of the underlying organizational behaviour and of diffusion processes in industry. For this reason, this paper shifts the emphasis away from the technical and economic features of energy efficiency measures and towards the communication and decision processes which determine the energy related behaviour of firms.

The objective is to apply a socio-economic approach in order to open the "black box", and learn about the people who act within the enterprise, their perception of problems and their mechanisms for generating and executing decisions. In the following I use "socio-economic approach" as a general, exemplary term for describing interdisciplinary research on organizational behaviour, combining aspects of business science, organization research, social-psychology, innovation research, etc. The paper aims to answer the questions:

1. What can be learned from socio-economic research about the implementation of efficiency measures in firms (sections 2 and 3)?
2. Can findings from socio-economic research be used to improve engineering-economic approaches by helping to give a broader understanding of transaction costs (section 4)?
3. What are the original contributions of socio-economic research to energy policy analysis (section 5)?

2. Opening the Black Box - A Closer Look at the Adoption of Energy Efficiency Solutions

This paper draws on a recent qualitative empirical study on the implementation of energy efficiency measures in small and medium-sized industrial enterprises (SMEs) (InterSEE, 1998; Ramesohl, 1998). To a large extent, the findings and conclusions derived have been supported by comparable research on the adoption of energy efficiency technologies in larger firms (TTI, 1998) and the implementation of energy efficiency activities in public institutions (IfP/ISI/WI, 1997, 1999). Although the studies covered rather heterogeneous institutions, it was possible to identify typical similarities and behaviour patterns. For the purposes of this paper, I would like to highlight five essential observations:

2.1. ECONOMIC DECISION IS ONLY ONE STEP IN A MORE COMPLEX PROCESS OF ADOPTION

By describing adoption as an act of economic decision, micro-economic theory dramatically simplifies the adoption of new technologies and organizational solutions. Empirical findings have clearly indicated that the moment of the investment decision represents just one stage among others which all together form a complex process of communication and interaction within an organization (Fig. 2). Accordingingly, the outcome of the adoption process is not exclusively determined by pure economic cost-benefit features of the conservation technology; social parameters and interrelations from process stages both before and after the investment decision are also of great importance for the final result. For this reason, the economic decision cannot be seen as an isolated event but has to be interpreted bearing in mind the context of the process. The following points will illustrate some of these aspects and interdependencies in greater detail.

Figure 2: The process of adoption and essential parameters and interdependencies

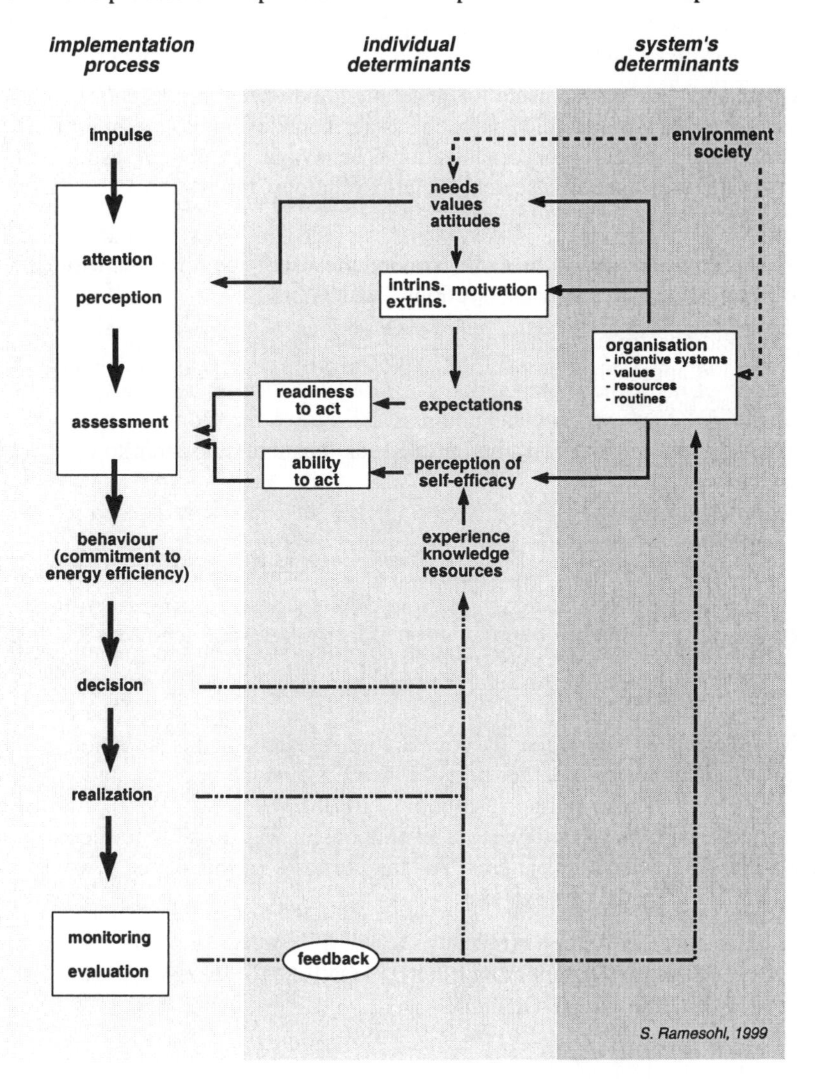

2.2. ENERGY ISSUES ARE USUALLY NEGLECTED AND NEED DISTINCT IMPULSES TO GET BACK ON THE AGENDA

Engineering-economic analyses usually start from the underlying assumption that industrial firms are actively searching for profitable investment opportunities. Companies are supposed to know already that they can adopt a particular promising option. For very energy-intensive processes in companies with their own experts this might be true. In many industrial sectors and most service companies and public institutions, however, energy represents only a minor share of total costs (on average 1-3%). It is not of any vital importance for business performance and the topic receives only little attention from the firm's executives. Energy efficiency options get low priority when competing with other internal investment opportunities, and they are not part of the management's daily considerations. Therefore, a theoretically sound economic decision in terms of a deliberate and active search for cost minimizing opportunities, a comprehensive specification of alternatives, and a proper cost/benefit assessment does not take place. In daily business practice, firms tend to handle energy issues on the side by standard procedures, routines and habitual decisions. Apart from energy intensive production processes, energy consumption is rarely monitored or systematically evaluated; replacement investments are often undertaken without considering alternative options. Moreover, in many SMEs, it is the electrician or plumber who determines the state of the firm's energy system – regardless of his or her competence. As a consequence, many firms stick to traditional but inadequate energy technologies, miss investment opportunities and maintain cooperations with unqualified but long standing market partners.

In nearly all cases, we observed that triggering impulses were needed to overcome this inertia and to initiate the adoption of a new energy efficiency solution by bringing energy back in mind as a variable for technical and economic optimization. Looking at our empirical experiences, four different areas of triggering impulses can be distinguished: **internal problems** force energy relevant action by the firm, e.g. when a machine breaks down and needs to be replaced. **External pressure**, too, enforces firm action e.g. by efficiency standards or environmental regulation. In both situations, the impulse breaks routines and creates an unavoidable pressure to act. **Internal opportunities** such as the move to a new location and **external incentives** (programmes, awards etc.) provide a suitable occasion to think about efficiency measures and to reduce the costs of implementation by synergies with measures which are being undertaken anyway or by external support (subsidies, etc.). Generally speaking, triggering impulses induce a re-allocation of attention so that the firm actors are – at least for a moment – attracted to the topic of energy. In order to result in concrete energy efficiency measures, however, the impulses need to be perceived as a promising opportunity for individual and collective action.

2.3. ENERGY EFFICIENCY NEEDS TO BE PERCEIVED AS A PROMISING OPPORTUNITY FOR INDIVIDUAL AND COLLECTIVE ACTION

At this very early stage of initiation, the first perception and mobilization of actors in a firm determines whether the process will continue or stop (risk of lost opportunity to act), and whether the most energy efficient solution will be taken or not (risk of lost opportunity to optimize). In this regard, two aspects are of especial relevance (Fig. 2):

As already known from social-psychology, economic variables are only one factor among others to explain individual motivation, allocation of attention and actors' perception of an opportunity to act. Economic actors do not live in a detached universe; besides expectations concerning the economic benefits of action, **individual readiness to act** is affected by underlying personal needs, values and attitudes (Bandura, 1986; Banks, 1999). Moreover, the individual sphere is embedded in the organizational context, and the determinants of individual behaviour are influenced by the common social background, the corporate culture and its related values, incentive systems, prevailing management styles, etc. Especially with regard to entrepreneurs from SMEs, for example, we observed strong links to the local and regional community and related social responsibility as driving forces for an engagement in energy efficiency.

In addition to their readiness to act, actors need to perceive the feeling of **ability to act**, i.e. they need to feel able to handle the tasks envisaged and to cope with possible problems and challenges[6]. Obviously, energy related knowledge and experiences as well as access to sufficient resources support this perception. The organizational context contributes to this feeling e.g. in terms of a management culture which is open to risk and which honours innovation and self-initiative by staff members.

2.4. EFFICIENCY PROJECTS RISK BEING TURNED DOWN DUE TO INSUFFICIENT DECISION PREPARATION

In micro-economic theory, firm actors live in a world of well-defined decision problems to choose from. Traditional analysis concentrates on the final investment decision, but the question of where decision alternatives come from is rarely tackled. In real-world business practice, however, the identification and specification of the original problem, its relevant parameters, appropriate decision alternatives and their consequences represent a challenging task of the pre-decision stage[7]. In the context of an energy efficiency project, the initial idea for an energy efficiency measure needs to pass through various steps of decision preparation such as data collection, auditing, detailed analyses and technical/organizational planning (Fig. 3). Depending on the complexity of the tasks, this planning requires work time, financial re-

6 In social-psychology, this aspect is discussed within the concept of self-efficacy, cf. Bandura 1986.

7 Cf. Hausschildt 1993, 108ff.,158ff.

sources and maybe even costly external support, and therefore demands pre-decisions by the executive(s) in charge.

Figure 3: The interdependency of technical planning and social interaction during the decision preparation phase

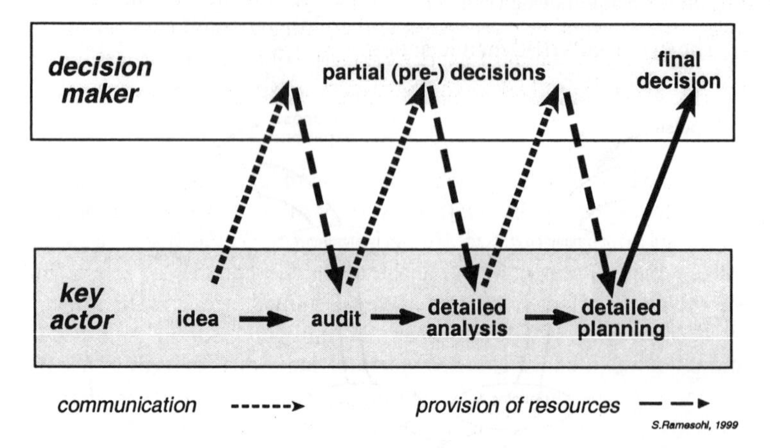

At this stage, empirical evidence points out the problem that if management refuses to release funds for the preparatory work, even highly profitable projects risk being turned down on the base of poor information, wrong expectations, insufficient decision criteria, unreflective routine behaviour, rules of thumb, etc.

Thus the preparation phase does not only cover the technical and economic analysis, the planning and the assessment of the measure in question. Simultaneously, a process of communication and social interaction takes place which is directed towards getting management support for the preparatory tasks. In this respect, empirical observations emphasize the important role of one or more **key actors** who feel personally responsible for the project, and show a strong commitment to pushing the idea. The key actors, usually members of the technical middle management, present the project to the decision-makers, defend it against criticism, and fight for the resources that are needed for continuation of the project. Normally, later on these key actors are in charge of the execution of the decision or are at least involved in the coordination of the realization process[8].

2.5. FEEDBACK LOOPS AND LEARNING EFFECTS CONTRIBUTE TO SELF-DYNAMIC ACTION

In engineering-economic analyses, each decision is usually considered as an isolated act on its own. Time does not matter, and from the analytical viewpoint it makes little difference whether the firm is undertaking energy efficiency measures for the

8 These findings are supported by innovations research which emphasizes the important role of "process champions" or "promotors" in pushing and coordinating an innovation project, cf. Witte 1973; Chakrabarti 1974; Howell, Higgins 1990; Gemünden, Walther 1995; Hausschildt, Schewe 1997.

first or for the tenth time[9]. However, in real life there is never a "tabula rasa". Every action depends on its historical context and the given consequences of previous activities, and in turn, it influences the preconditions for subsequent projects. The end-point of one project represents the starting point of the next, so that each project is linked to the subsequent one (Fig. 4).

Figure 4: The cycles of self-dynamic action

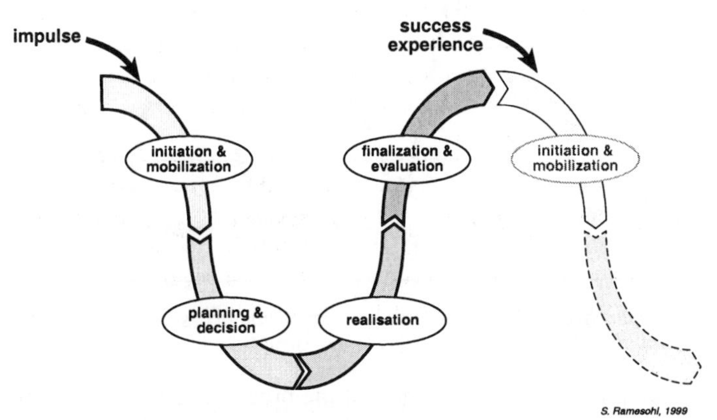

From this viewpoint, continuous successful activity appears as a series of cyclic or spiral-like process loops which constitute a learning and development process. In the long run, the firm can move towards a more energy efficient status and culture. At the same time, however, failure and disappointment can induce destructive dynamics which cause the preconditions for efficiency action to deteriorate.

Any energy policy intervention which aims to enhance self-supporting energy efficiency action needs to address and enhance this learning process. For this reason, explicit consideration of the dynamic interdependencies and feedback loops within the adoption process is of special importance in order to understand the nature of long-lasting change and explore the possibilities for an active stimulation of self-dynamic development (cf. Fig. 2):

1. **Motivation, expectations and the readiness to act** strongly depend on individual experiences from previous actions. Positive experiences facilitate the mobilization of actors, whereas negative results can inhibit activities for years. In this regard, actors who have been involved in an energy efficiency measure need to get a feeling of success in order to build up positive experiences. Explicit monitoring and feedback of project results, therefore, contribute to stabilizing and improving motivation and commitment [10].

---

9 The omission of time interdependencies and history is one fundamental strand of criticism of neo-classical economic theory by evolutionary economics, cf. Nelson, Winter 1982; Boulding 1990; Dosi 1991; Dosi, Nelson 1994.

10 The possibility to win an energy award, for example, may hardly motivate firms *ex-ante* to undertake conservation projects. However, the empirical evidence suggests that an *ex-post* recognition of action seems to have a strong influence in keeping the momentum and maintaining the motivation to act.

2. During the course of implementation audits, planning data and monitoring experiences enlarge the pool of energy related information in the firm and thus improve the **background for further project planning, evaluation, and decision preparation.**
3. Actors involved during the implementation of energy efficiency measures gain direct experience of the new energy efficiency technology. This contributes to building up energy related personnel skills, know-how and tacit knowledge. In addition, actors learn about communication with management, collaboration with colleagues in the firm and cooperation with external partners. All these experiences – which can be enhanced by specific training and qualification - contribute to an **accumulation of technical and social competence,** and, therefore, increase the staff's ability to act.
4. In addition to technical investments, energy efficiency projects can induce a **change in organizational practices and structures**. A shift in responsibilities, new data collection, additional monitoring and energy management devices, as well as altered routines and procedures (e.g. with regard to procurement) contribute to a modification of the organizational context. This facilitates the adoption of energy efficiency technology in the future.

The aspects mentioned above can be summarized as energy efficiency related learning on the individual and collective level which requires working communication and feedback loops. At this point it is important to note that the socio-economic concept of learning goes far beyond the micro-economic representation of knowledge. In engineering-economic analyses, knowledge is often understood as the access to and the distribution of exchangeable information. To be equipped with information, however, does not tell anything about the capacity to use the information meaningfully (Nelson and Winter, 1982).

And another remark needs to be made: usually, learning is applied to an improvement of abilities and performance in relation to a certain task, i.e. learning in the sense of doing things differently. According to Argyris (1977), this type of learning can be summarized as **single loop learning**. At the same time, however, actors get the chance to question and uncover their own hidden mechanism of action, i.e. to identify underlying paradigms, objectives and rules of behaviour as preconditions for various actions. This **double loop learning** corresponds to learning on a meta-level through gradual changing of the organizational context in terms of modification of organizational structures and incentives, adaptation of routines and processes, changing habits and even an evolution of new values and corporate culture.

3. Dimensions of the Implementation of Energy Efficiency Solutions

The brief discussion of process steps, behavioural determinants and dynamic interdependencies illustrates the contribution that socio-economic research can make to a more detailed understanding of adoption and diffusion in industry. Without questioning the relevance of rational economic behaviour for industrial decision-making, new emphasis is given to the dynamic and social context of investment behaviour.

This represents a shift in analytical perspective. Compared with engineering-economic analyses and their micro-economic foundation, the focus is no longer put on an impersonal act of decision but on a **social process of problem identification, problem definition, and problem solving**. A new dimension of thinking in terms of dynamic, social systems has been introduced[11].

Accordingly, industrial behaviour does not only comprise the choice of a technology but covers all aspects of the **implementation of a solution** in the sense of a package of technical and organizational measures. Different actors contribute to this joint endeavour and perform various actions which are needed to start the process, to initiate and perform the technical planning, to convince top management, to coordinate external partners, to execute the decision, to qualify and supervise the staff, to monitor the results etc. Taking all these contributions into account, the implementation of an energy efficiency solution is characterized by the **three complementary dimensions of technology provision, techno-economic transaction and social interaction,** which cannot be analysed separately without missing decisive parameters and interdependencies (Fig. 5).

Figure 5: The three dimensions of an energy efficiency solution

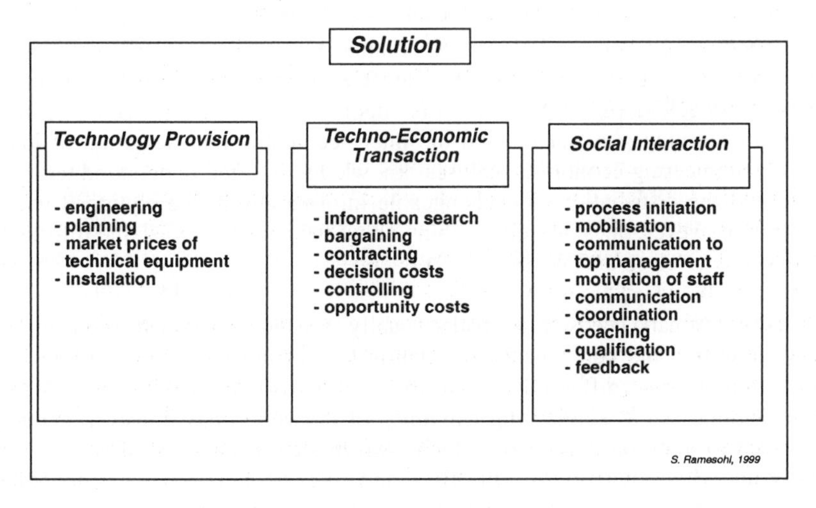

What can be learned from these observations? On the one hand, the dimensions of techno-economic transactions and social interactions cover the totality of non-material contributions to the process of adoption. Some of these elements are already accounted for when calculating technology costs (e.g. engineering or planning) but other aspects have hardly been recognized so far as distinct features or cost parameters of the solution, especially in the field of communication and social interaction. Hence, new insights into the nature of transaction costs can be derived (see section 4). On the other hand, the analysis reveals important features of change pro-

---

11 Senge calls this the "fifth discipline" " which forms a basic prerequisite for the creation of learning organizations" (Senge 1994).

cesses in organizations which cannot be operationalized by transaction cost economics. Here, an original contribution by socio-economic research to energy policy analysis can be identified (section 5).

4. New Insights into the Nature of Transaction Costs

Techno-economic modellers tend to expect socio-economic research to provide better cost data. Social science should translate – at least partially – the identified domain of social interactions and communicative settings into economic terminology, and operationalize the insights for quantitative energy policy analysis. The hope is that an enlargement of the cost assessment might serve as a bridge between techno-economic approaches and social science, thus opening up a possibility to introduce behavioural aspects into engineering-economic policy analyses. From this viewpoint, all aspects identified need to be prepared for treatment as individual cost categories which can then be taken into account when assessing the economic efficiency of energy efficiency measures and CO_2 mitigation at the level of firms. So the relevant questions are:

What are the concrete implications for the improvement and development of engineering-economic analyses? Is it possible to translate socio-economic findings into economic cost categories and integrate them smoothly into existing CSCs, cost assessments and policy analyses?

In my opinion, the answers are ambiguous: on the one hand, socio-economic research reveals forgotten costs and efforts which occur during the process of adoption and implementation of an energy efficiency measure. By broadening the focus and improving the understanding of the communicative and organizational aspects of energy efficiency projects, new cost factors come into consideration which have not been sufficiently represented in engineering-economic analyses so far[12]. In this respect, a more comprehensive picture concerning the transaction costs or so called "hidden costs of energy efficiency" emerges, and the extension and re-interpretation of cost factors may lead to modified supply curves and changed cost relations between technologies. Cost analyses which neglect these socio-economic factors risk giving a distorted picture of the cost effectiveness of energy efficiency technologies.

On the other hand, there are clear limits to a quantification of social parameters in terms of transaction costs. Empirical observations leave no doubt about the fact that the interaction and exchange of services with external partners is of great importance for a successful adoption of energy efficiency technologies. External actors play an important role and they often give the decisive kick to start action, provide valuable information on possible technologies and support the internal key actor(s) in their tasks of coordinating the process. In many cases, certain external actors form a close team together with the internal key actor(s) and work as crucial driving force for

[12] The structure and size of time allocation, for example, might serve as a first proxy for necessary actions and related implementation costs (cf. Witte 1973). However, at the current stage little empirical evidence is available in this regard, see Hein, Blok, 1995.

success. However, little can be said *ex-ante* about which persons or combinations are going to be most cost-effective in the given context, or how to create these constellations. Moreover, due to the manifold institutional settings, market structures and social constellations in place, clear-cut information about size and direction of costs can hardly be derived. Touching the core question "make-or-buy" of transaction-cost economics, an ambiguous trade-off can be observed between reducing the internal implementation effort on the one hand, and rising costs for coordinating and controlling the external contributors and third-party contractors on the other hand[13].

At the moment, the discussion on hidden costs and benefits, on their nature and size as well as on adequate methodologies for quantification is still at its beginning. If they are included at all, hidden costs are insufficiently represented in engineering economic analyses, e.g. when modelled as a fixed surcharge of X % on all investment costs. The following methodological aspects still need to be elaborated (Ostertag, 1999):

1. When integrating further cost categories, a **full impact assessment** has to take place. This includes not only a complete assessment of hidden costs of the new alternative but also a complete assessment of the status quo solution, in order to avoid unjustified bias and discrimination against the new option. Furthermore, hidden costs should be enlarged to include transaction benefits and hidden profits in order to give a correct assessment of net benefits. To assess the net benefits, cost estimates should be based on life cycle assessment in order to take into account both the up-front costs and the follow-up expenses during operation.
2. As depicted in section 2, energy efficiency measures often represent a change of prevailing routines and practices in the firm which creates additional costs such as market inquiries, search for new partners, motivation of staff, etc. These costs only occur, however, at the beginning and may be substantially reduced with stabilization of the new practice, e.g. when a new supplier has been found. For this reason, it is necessary that the assessment of costs should account for **dynamic aspects** such as learning, economies of scale and scope, etc.
3. Policy programmes are dedicated to supporting the implementation process and, therefore, to reducing hidden costs. When assessing the economic efficiency of policy intervention, analysis has to take care to **avoid a double counting of transaction costs** both as implementation costs in the CSC, and a second time as programme costs.

As the most important aspect, however, I should like to emphasise that the use of the transaction costs concept should not be limited only to generating new explanations for decision-making. The extension of cost assessment opens up more possibilities

13 In this regard, it would be interesting to discuss Williamson's notion of asset specificity in more detail, in order to gain better insight into the question whether process contributions are best provided by the market or by the organization (Williamson 1985).

than a simple tautological exercise in ex-post rationalization of firm behaviour[14]. The socio-economic view of transaction costs as policy variables provides a starting point not only for more comprehensive analysis of impact factors and implementation efforts, but also for the identification of policy options to minimize the necessary efforts and to enhance energy savings.

5. The Original Contribution of Socio-Economic Research to the Policy Domain of Technical Progress and Diffusion

Socio-economic research can be used to a certain extent to improve engineering-economic analyses. The primary benefit, however, should be seen in an original input to the policy challenge of inducing technical progress, technology diffusion and market transformation processes in dynamic systems.

In order to understand the full range of possibilities for enhancing the diffusion of energy efficiency solutions, the identified learning effects are of special importance. In recent years, particular efforts have been undertaken to analyse the impact of learning, experience curves and technical progress on technology costs (Messner, 1997; Neij; 1998; Wene, 1998). For the main part, however, these attempts have concentrated on the production costs and market prices of energy technologies such as PV or wind turbines. Industrial organizations have primarily been seen as producers and sellers of technologies, whereas the procurement of goods and services by institutional buyers is still neglected[15]. From a socio-economic perspective, however, this is only half the story. As we have seen, learning effects and experience curves can be observed on the demand side, too, and these have a significant influence on the process of adoption. Rather than examining only the technical content of diffusion, more attention needs to be given to the social and organizational context of diffusion in terms of sustained change in behaviour, organizational routines and production structures on the side of the potential adopters. Like technology learning on the supply side, social and organisational **adoption learning on the demand side** represent a promising domain for research and energy policy-making. The following two core areas for action can be identified:

1. **Enhancing demand-side adoption learning** by creating a general culture of learning, process thinking, and continuous improvement processes at the level of firms. The evolution of feedback loops, monitoring and evaluation of data e.g. by energy management systems, as well as having an effect on the organizational culture, contributes both to first and second level learning. In this respect, already-existing theories and approaches that deal with change in organizations (e.g. with the "learning organization") are a promising source of inspiration and

---

14 "Transaction costs have a well-deserved bad name as a theoretical device..., [partly] because there is a suspicion that almost anything can be rationalized by invoking suitably specified transaction costs" (Fischer, quoted from Williamson 1996).

15 This perspective reflects the conceptual heritage of the "Theory of Industrial Organizations" which emphasizes the competitive and innovative aspects of industrial suppliers (Kaufer 1980, Tirole 1988).

empirical evidence[16]. Enhancement of the still insufficient exchange between the various areas of research and policy-making will significantly increase theoretical and empirical knowledge about organizations. In addition, it will be possible to benefit from synergies between the different policy areas of energy, environment, regional economic development, R&D and innovation, professional qualification and training, etc.

2. **Enhancing processes of change in socio-economic systems** by creating and increasing the probability of interaction. This requires an appropriate milieu for social learning (Bandura 1986) and a smooth exchange of information among the various partners and market actors.

With regard to the latter aspect, markets are commonly perceived as the most appropriate system to provide and coordinate the necessary interactions and exchanges of services. The domain of energy efficiency, however, is characterized by the fact that an existing efficiency solution does not automatically meet with a corresponding demand. Due to the low priority of energy issues, many consumers simply ignore profitable options for reducing energy use. Contrary to neo-classical thinking, at the same time it is impossible for a third agent to take advantage of the untapped potential. In this regard, energy efficiency markets can be considered to be "in fact almost perfectly uncompetitive" (Johnson, 1994). For this reason policy analysis is required, not only to investigate barriers and distortions of existing markets, but also to provide insights into the possibilities to create new business fields and build up market relations from scratch. The task is to contribute to the formation of markets which are still at a pre-competitive stage.

In addition, more knowledge is required in order to specify the particular kind of services and energy efficiency tools which, although needed for organisational change, are likely to remain untradeable under professional conditions, e.g. motivation and mobilization measures or coaching of project partners. In order to stimulate and accelerate energy efficiency activities these services will then have to be provided by non-market forces such as public agencies.

Finally, when re-directing attention from technologies to the context of adoption, process interactions and cooperations gain importance. Rather than looking at isolated actors with different backgrounds, complete networks and groups such as supply chains have to be put at the centre of policy interest, and addressed by comprehensive approaches.

Summing up, a **socio-economic systems analysis** is required which investigates opportunities to induce or to change network relations, to disseminate competence and motivation within and between organizations, and to seed new clusters of activity. In other words, it is needed to create the communication structures in social

---

16 Interestingly, the economic perspective of energy research is dominated by neoclassical microeconomics, whereas the rich treasury of business and management science remains almost completely ignored, see for example Senge (1994).

systems which are the mandatory precondition for the diffusion of innovation (Rogers, 1995).

6. Conclusions

Compared to traditional barrier analyses, the approach presented here has generated a new quality of insights into implementation processes. It has been possible to describe a broad range of factors influencing the adoption of energy efficiency solutions, from the viewpoint of concrete realisation. Especially social, psychological and organizational determinants of energy efficiency implementation – which are often neglected and underestimated in their impact – have been identified and translated into policy recommendations. In addition, the analysis of positive examples has made it possible to study the dynamic nature of long-term success, e.g. the prerequisites for the development of an "energy efficiency culture" in SMEs, starting with simple measures and continuing with relatively complex activities such as eco-management. Rather than aiming at a mechanical and linear enforcement of isolated instrument-impact schemes, therefore, effective policy-making should strive for a simultaneous and synergetic impact on three levels:

- overcoming the target group's lack of awareness and the inertia stemming from prevailing routines and the low priority assigned to energy,
- changing energy related routines and behaviour through internal learning and orga- nizational changes (establishment of energy efficiency infrastructure and culture),
- formation and support of energy efficiency networks and energy service market relations.

These new priorities will have considerable implications for programme design and management as well as for policy evaluation (see. Fig. 6). The focus, traditionally put on single instruments which are targeted at the economic decision-making, needs to be shifted towards policy mixes, aiming to induce and direct energy efficiency-related co-operation between various actors from various target groups. In this respect, traditional evaluation criteria have to be interpreted in a new way. Especially the notion of dynamic efficiency – usually applied to technical progress – has to be enlarged and strengthened. The evaluation procedure also has to look for learning effects, long-term changes of preferences and new codes of conduct, as well as indirect effects which seem likely to foster the implementation of efficiency measures in the future. This type of impact is often connected with 'soft' measures such as network formation, education, professional training etc. – areas in which quantitative data is hard to generate.

Figure 6: Shift of energy policy evaluation paradigms

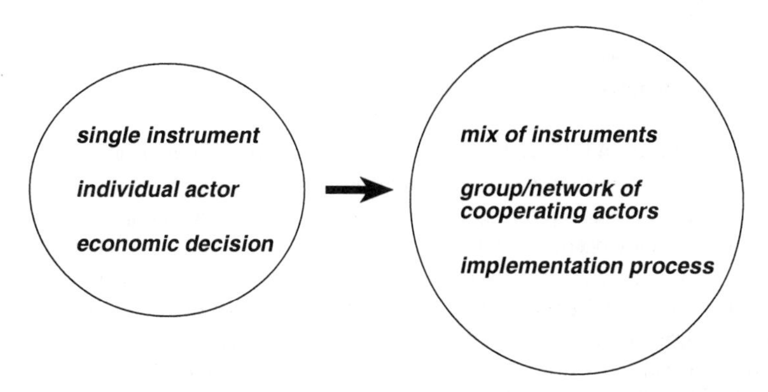

Furthermore, traditional evaluation tends to restrict itself to partial analyses of isolated instruments rather than trying to grasp the complex interactions generated by policy mixes. However, this evasion – motivated by methodological simplification and analytical pragmatism – incorporates the risk of falling short in explaining real world interdependencies, and thus providing inappropriate advice for energy and climate policy. The development of suitable methodologies and criteria for the assessment of energy policy mixes still represents a scientific challenge and a promising area for socio-economic energy research.

7. References

Argyris, C., 1977: Double loop learning in organizations. *Harvard Business Review*, Sept./Oct. 1977, pp. 115-125.

Ayres, R., 1994: On Economic Disequilibrium and Free Lunch, *Environmental and Resource Economics*, 4(5), pp. 435-479.

Bandura, A., 1986: *Social Foundations of Thought and Action - A Social Cognitive Theory*, Englewood Cliffs.

Banks, N., 1999: Causal models of household decisions to choose the energy efficient alternative: the role of values, knowledge, attitudes and identity. In: *Energy Efficiency and CO_2 reduction: the Dimensions of the Social Challenge*, Proceedings of the 1999 Summer Study of the European Council for an Energy Efficient Economy, Part 2, Paper (III.24), ADEME (ed.) Paris.

Boulding, K., 1991: What is evolutionary economics? *Journal of Evolutionary Economics*, 1(1), pp. 9-17.

Carlsmith, R., Chandler, W., McMahon, J., Santini, D., 1990: *Energy Efficiency: How Far Can We Go?* Report ORNL/TM-11441, Oak Ridge National Laboratory.

Chakrabarti, A., 1974: The Role of Champions in Product Innovation, *California Management Review*, No. 7, pp. 58-62.

Cyert, R. and J. March, 1992: *A Behavioural Theory of the Firm*, 2nd ed., Cambridge/Mass., Oxford.

DeCanio, S., 1993: Barriers within Firms to Energy-Efficient Investments, *Energy Policy*, 21(9), pp. 906-914.

DeCanio, S., 1998: The energy efficiency paradox: bureaucratic and organizational barriers to profitable energy-saving investments, *Energy Policy*, 26(5), pp. 441-454.

Dosi, G., 1991: Some thoughts on the promises, challenges and dangers of an "evolutionary perspective" in economics, *Journal of Evolutionary Economics*, 1(1), pp. 5-7.

Dosi, G. and R. Nelson, 1994: An introduction to evolutionary theories in economics, *Journal of Evolutionary Economics*, 4(4), pp. 153-172.

European Commission, 1998: *Energy Efficiency in the European Community - Towards a Strategy for the Rational Use of Energy*, Communication of the Commission, COM (98)246, Brussels.

Fisher, A. and M. Rothkopf, 1989: Market failure and energy policy - A rational for selective conservation, *Energy Policy*, 17(8), pp. 397-406.

Gemünden, H. and A. Walter, 1995: Der Beziehungspromotor – Schlüsselperson für inter-organisationale Innovationsprozesse, *Zeitschrift für Betriebswirtschaft*, 65(9), pp. 971-986.

Gillissen, M. and H. Opschoor, 1994: *Energy Conservation and Investment Behaviour: An Empirical Analysis of Influencial Factors and Attitudes*, Serie Research Memoranda, Research Memorandum 1994-31, Faculteit der Economische Wetenschappen en Econometrie, Vrije Universiteit Amsterdam, Amsterdam.

Gillissen, M., 1994: *Energy Conservation Investments - A Rational Decision?* Serie Research Memoranda, Research Memorandum 1994-33, Faculteit der Economische Wetenschappen en Econometrie, Vrije Universiteit Amsterdam, Amsterdam.

Hasset, K. and G. Metcalf, 1993: Energy Conservation Investment - Do Consumer Discount the Future Correctly? *Energy Policy*, 21(6), pp. 710-716.

Hasset, K. and G. Metcalf, 1996: Can Irreversibility Explain the Slow Diffusion of Energy Saving Technologies? *Energy Policy*, 24(1), pp. 7-8.

Hauschildt, J., 1993: *Innovationsmanagement*, München.

Hauschildt, J. and G. Schewe., 1997: Gatekeeper und Promotoren: Schlüsselpersonen in Innovationsprozessen in Statischer und Dynamischer Perspektive, *Die Betriebswirtschaft*, 57(4), 506-515.

Hein, L. and K. Blok, 1995: Transaction Costs of Energy Efficiency Improvement. In: *Sustainability and the Reinvention of Government - A Challenge for Energy Efficiency*, Proceedings of the 1995 ECEEE Summer Study. The European Council for an Energy Efficient Economy, NUTEK (ed.) Stockholm.

Howarth, R. and B. Andersson, 1993: Market Barriers to Energy Efficiency, *Energy Policy*, 21(10), pp. 262-272.

Howell, J. and C. Higgins, 1990: Champions of Technological Innovation, *Administrative Science Quarterly*, No. 35 (1990), pp. 317-341.

IfP/ISI/WI, 1997: *"Interdisziplinäre Analyse der Umsetzungschancen einer Energiespar- und Klimaschutzpolitik: Hemmende und fördernde Bedingungen der rationellen Energienutzung für private Haushalte und ihr Akteursumfeld aus ökonomischer und sozialpsychologischer Perspektive"*, Projekt Klimaschutz/Institut für Psychologie, Fraunhofer Institut für Systemtechnik und Innovationsforschung, Wuppertal Institut, Kiel, Karlsruhe, Wuppertal.

IfP/ISI/WI, 1999: *"Mobilisierungs- und Umsetzungskonzepte für verstärkte kommunale Energiespar- und Klimaschutzaktivitäten"*, Projekt Klimaschutz/Institut für Psychologie, Fraunhofer Institut für Systemtechnik und Innovationsforschung, Wuppertal Institut, Kiel, Karlsruhe, Wuppertal.

International Energy Agency (IEA) (eds.), 1999: *Technologies to reduce Greenhouse Gas Emissions - Engineering economic analyses of conserved energy and carbon*, Proceedings of an international workshop, Washington/D.C., 5-7.5.1999, International Energy Agency, Paris.

InterSEE, 1998: *Interdisciplinary Analysis of Successful Implementation of Energy Efficiency in Industry, Commerce and Service.* Wuppertal Institut für Klima Umwelt Energie, AKF-Institute for Local Government Studies, Energieverwertungsagentur, Fraunhofer Institut für Systemtechnik und Innovationsforschung, Projekt Klimaschutz am Institut für Psychologie der Universität Kiel, Amstein&Walthert, Bush Energie, Wuppertal, Kopenhagen, Wien, Karlsruhe, Kiel.

Jaffe, A. and R. Stavins, 1994: The Energy Efficiency Gap. What Does It Mean? *Energy Policy,* 22(10) pp. 804-810.

Jochem, E. and E. Gruber, 1990: Obstacles to the Rational Use of Electricity and Measures to Alleviate Them, *Energy Policy,* 18(5), pp. 340-349.

Johnson, B., 1994: Modelling Energy Technology Choices - Which Investment Analysis Tools Are Appropriate? *Energy Policy*, 22 (10), pp. 877-883.

Kaufer, E., 1980: *Industrieökonomik - Eine Einführung in die Wettbewerbstheorie*, München.

Koomey, J. and A. Sanstad, 1994: Technical Evidence for Assessing the Performance of Markets Affecting Energy Efficiency, *Energy Policy*, 22 (10), pp. 826-832.

March, J., 1994: *A primer on decision making*, New York.

March, J. and H. Simon, 1993: *Organizations*, 2nd ed., Cambridge/Mass., Oxford.

Messner, S., 1997: Endogenized Technological Learning in Energy Systems Models, *Journal of Evolutionary Economics*, 7(7), pp. 291-313.

Metcalf, G., 1994: Economics and Rational Conservation Policy, *Energy Policy*, 22 (10), pp. 819-825.

Neij, L., 1998: The Use of Experience Curves to Analyse Cost Reductions in Renewable Energy Technologies. In: *Technological Innovations, Sustainable Development and the Kyoto conference*, [Science Policy Research Unit (SPRU) (ed.)] ENER Bulletin 22.98, p.9, Brighton.

Nelson, R. and S. Winter, 1982: *An Evolutionary Theory of Economic Change*, Cambridge/Mass., London.

Ostertag, K., 1999: Transaction Costs of raising energy efficiency. In: *Technologies to reduce Greenhouse Gas Emissions - Engineering economic analyses of conserved energy and carbon*, [International Energy Agency (IEA)(eds.)], Proceedings of an international workshop, Washington/D.C., 5-7.5.1999, International Energy Agency, Paris.

Ramesohl, S., 1998: Successful Implementation of Energy Efficiency in Light Industry. In: *International Workshop on Industrial energy Efficiency Policies: Understanding Success and Failure*, [Lawrence Berkeley National Laboratory (ed.)], Proceedings of the international workshop of the International Network for Energy Demand Analysis in the Industrial Sector (INEDIS), Utrecht, 11./12.6.1998, LBNL-42368, Berkeley CA.

Rogers, E., 1995: *The diffusion of innovations*, 4th ed., New York.

Sachs, W., Loske, R., Linz, M., 1998: *Greening the North - A Post-Industrial Blueprint for Ecology and Equity*, London, New York.

Sanstad, A. and R. Howarth, 1994: "Normal" Markets, Market Imperfections and Energy Efficiency. *Energy Policy*, 22 (10), pp. 811-818.

Senge, P., 1994: *The Fifth Discipline – The Art and Practice of the Learning Organization*, New York et al.

Sutherland, R., 1991: Market Barriers to Energy-Efficiency Investments, *The Energy Journal*, 12(3), pp. 15-35.

Sutherland, R., 1994: Energy Efficiency or the Efficient Use of Energy Resources? *Energy Sources*, 16, pp. 257-268.

Sutherland, R., 1996: The economics of energy conservation policy, *Energy Policy*, 24 (10), pp. 361-370.

The Tavistock Institute (TTI) (ed.), 1998: "*Understanding the take up of advanced energy technologies*". Final Dissemination Report. The Tavistock Institute, AROC, Danish Technological Institute, Free University Amsterdam, London.

Tirole, J., 1988: *The Theory of Industrial Organization*, Boston.

Verdonck, P., J. Couder, A. Verbruggen, 1998: The Use of Conservation Supply Curves in Policy Making. In: *Technological Innovations, Sustainable Development and the Kyoto conference*, [Science Policy Research Unit (SPRU) (ed.)] ENER Bulletin 22.98, p.19, Brighton.

Weber, L., 1997: Some Reflections on Barriers to the Efficient Use of Energy, *Energy Policy*, 25(10), pp. 833-835.

Wene, C., 1998: *Energy Technology Price Trends and Learning*, Note by the IEA Secretariat [IEA/CERT(98)24], International Energy Agency, Committee on energy research and technology, Paris.

Williamson, O., 1985: *The Economic Institutions of Capitalism*, New York.

Williamson, O., 1996: Efficiency, Power, Authority and Economic Organizations. In: *Transaction Costs and Beyond* [Groenewegen, J. (ed.)], Boston et al., pp. 11-42.

Witte, E., 1973: *Organisation für Innovationsentscheidungen: Das Promotorenmodell*, Göttingen.

Acknowledgement

The author thankfully acknowledges the funding of the underlying research by the European Commission DGXII under the VAIE project (JOS3-97-0021) and the InterSEE project (JOS3-95-0009) and the fruitful exchange and collaboration with the involved partners.

Motivation and decision criteria for energy efficiency in private households, companies and administrations in Russia

Inna Gritsevich

1. Introduction

Russia is one of the greatest world energy producers and consumers and hence a big emitter of greenhouse gases (GHG). More than 90% of Russia's GHG emissions are energy-related to a great extent. The energy intensity of the Russian economy was 2 to 3 times greater than that of the USA or Western Europe by the beginning of the 1990s (RF Ministry of Fuel and Energy, 1998). During the last decade, the energy intensity of the GDP even increased by more than 30%, mainly due to the economic decline of the Russian economy and the increasing motorization and use of electrical appliances in private households. Improvement in the energy efficiency of the Russian economy is a precondition for its competitiveness in world markets, and for future economic growth. Thus the energy sector and energy efficiency are very important, not only in connection with the climate change problem, but also as a basic economic problem in Russia. This problem is recognized by the majority of policymakers and governmental officials, thus improving energy efficiency was accorded the highest priority in the national Energy Strategy (IEA/OECD, 1995), and the Federal Programme of Energy Saving (RF Ministry of Fuel and Energy, 1998) has been developed and implemented since 1993, though not very successfully. However, neither of these documents pay any attention to psychological, social and cultural aspects or the related conceptual frameworks of energy saving, nor do they consider public awareness or personal motivations to save energy as important means to reach their goals.

Several times over the last 20 years, there have been efforts to increase public awareness of the energy efficiency problem. Public TV educated people on energy saving at prime time, but there were no studies of the effects of this promotion of energy saving. Introduction of some primary information on energy saving into educational programmes for schoolchildren was discussed.

In April 1998 CENEf performed a research survey on climate awareness in Russia, at the request of the UNEP (UNEP, 1998). Among other tasks this included interviewing five persons who are actively involved in climate change issues and are highly knowledgeable about these issues. They represented governmental officials, environmental NGOs, business leaders and officials from multilateral organizations. Four of them mentioned the energy sector, and energy efficiency in particular, as the most significant issues relating to global climate change in the Russian Federation.

No special sociological surveys or studies of behavioural aspects of energy saving and climate mitigation were performed in the last decade. At the same time, more general analysis of decision-making in households, businesses and administration can help to some extent in understanding the situation. For this reason our conclu-

E. Jochem et al. (eds.), Society, Behaviour, and Climate Change Mitigation, 227–237.

sions on motivations for energy saving behaviour in private households, businesses and administrations with regard to energy efficiency in Russia, presented below, will be based on the following sources:

(i) Sociological surveys and studies;
(ii) Publications on energy saving activities of private companies and administrations;
(iii) Personal communications and discussion;
(iv) Official documents on minimal and average living standards and statistical data.

2. Basic motivations and rationales for energy saving

2.1. REGIONAL AND MUNICIPAL ADMINISTRATIONS

In Russia, just under 20% of fossil fuels are consumed in heat production for municipal heat supply, making a significant contribution to national greenhouse gases emissions. District heating is a typical municipal heat supply system in Russian cities. The efficiency of these systems in Russia is much lower (2-3 times) compared to their efficiency in Western Europe (Matrosov, 1997). Currently residential customers pay only a part of the costs of their heat and power supply, though the share of these payments in bills for utilities services has increased from 40% to 70%. Public funding for this sector amounts to 17.5 billion US$ annually. Elimination of the subsidies by 2005, and reform of the housing sector, were identified as the key to continuation of the transition to a market economy (RF Ministry of Fuel and Energy, 1998).

Municipalities and regional administrations have to spend hard money from their budgets (20-35%) to subsidize residential heat customers. They usually recognize the importance of the problem and are interested in improving the energy efficiency of district heating. Money saved as a result of the potential improvements and the reduction of heat subsidies could be used by the municipalities, for instance to finance local social programmes which are urgently needed, but also to reinvest into cogeneration substituting simple boiler systems or in modern heat distribution and insulation in buildings owned by the municipalities, as they cannot benefit from improved financial situations.

A number of barriers prevent municipalities from implementing potential energy savings in the residential sector. Deficits of finances in the local budgets as a result of poor tax collection, corruption, mutual debts and non-payments, lack of personal incentives for bureaucrats, lack of qualified staff and lack of information on energy-efficient technologies and management, etc., should be mentioned here. One more complication results from the lack of the economic incentives for heating companies owned by municipalities to save energy.

The energy bills of public services funded by federal and local administrations (armed forces, schools, hospitals, etc.) are paid from the corresponding budgets too. So local authorities are interested in their reduction. But under existing funding system and with frequent delays of money transfers and underpayments of these public institutions, their management has no incentives to save energy because these savings might be used by funding authorities as a reason to reduce general funding. Moreover, usually there are no energy managers on the staff of public institutions to design and implement energy saving measures, and no funds to pay external experts to do this. So in many cases where it is possible to improve poor indoor conditions without increasing heat and power bills, this opportunity is not used (Vorobiev, et al., 1998).

2.2. PRIVATE HOUSEHOLDS

Traditionally, in old Russia, where more than 80 per cent of the population lived in the countryside in one-family houses, energy was used in a very efficient way. With accelerated forced urbanization in the last century these traditions were lost. A new way of living in multi-storied standardized state owned dwellings, characterized by low energy prices and lack of meters for natural gas, water or heat consumption (only power consumption is measured, with very simplistic equipment) in the apartments, did not create new motivations for the rational use of energy or rationales for reasoning about it. Until now, the supply of energy to residential customers has been planned by local and regional administrations on the basis of the official consumer norms for basic products and services. These norms are low compared to Western standards, since they assume quite small sizes of the apartments, modest equipping of the households with electric appliances, priority for public transportation over private cars, etc.

Last year, the updated version of individual minimal social consumer norms for basic products and services (consumer basket) and the methodology for their estimation was adopted by the government (RF, 1999). It assumes the following norms for energy use or energy services per capita:

District heating – 28 GJ annually;
Hot and potable water consumption – 285 l daily;
Natural gas (for cooking) – 390 MJm^3 monthly;
Power consumption (for illumination and electric appliances) – 50 kWh monthly;
Transportation services – 442 trips annually.

Thus the yearly per capita energy consumption in the former Soviet Union, and now in Russia (about 175 GJ/cap. a in 1996 (IEA/OECD, 1998)), has been gained as a result of the State policy of indirect energy rationing, not as a result responsible personal choices.

With the beginning of transition to a market economy, in the situation of deep economic and structural crisis and instability over the whole last decade, a great part of the population faced a dramatic decline of their personal incomes, the danger of losing their job and becoming unemployed, partial destruction and insufficiency of social guarantees and insurance, etc. It was expected that the massive privatization of apartments, started in this decade, could give households economic incentives for energy saving. In reality, no reduction of energy consumption in the residential sector was observed. Technical barriers to energy saving in this sector are: lack of metering and controlling equipment, reflecting a long-held view that heat and fuel are public goods, lack of differentiated tariffs for electrical power, etc.

Recent sociological studies of decision-making and resource allocation by households have shown that a great part of the population has no clear idea of rational planning of their expenditure or prioritizing their demands, despite the need to use their quite low income very economically (Manning, N., et al., 2000). The reasons for this kind of behaviour are:

(i). High uncertainty regarding the current and future economic and social situation of the family, which makes people consider any planning senseless. This is in great contrast with the past stability and certainty of the personal life cycle in the former social system.

(ii). Lack of experience in making responsible choices in the past. Under the centrally planned command system, allocation of goods and services did not in fact depend much on personal decisions, but were determined by a person's social and professional position, with the well known ways to change the position.

(iii). Lack of experience in allocating limited financial resources. In the former quasi-monetary Soviet economy where a large role was played by non-monetary welfare, limited access to goods and services was a problem, and the limited money reserves of individual households played a smaller role. Now, with the start of transition to a market economy, the situation has changed dramatically and money has become the main constraint for every economic agent. It takes time for a great part of the population, administrations and managers to adapt to this new environment.

Lack of access to some consumer goods for decades made them over-evaluated. The private car can be taken as a typical example. It was prestigious to have a car in the former Soviet Union and there was no free market for them. So in the last decade, the number of private cars was increasing despite the general decrease in personal incomes. People tried to satisfy their delayed desire and demonstrate their capabilities in this way. At the same time, the greatest demand has been for cheaper, outdated Russian-made and Western second-hand cars which have a low energy efficiency and are environmentally unfriendly.

So people need to be educated to make economically rational decisions, to plan their budgets and their spending. To help people purchasing refrigerators to make rational choices, CENEf – for example – developed the first programme in Russia on energy efficiency labelling for refrigerators: "ENERGOCOMPASS". The "ENERGOCOMPASS" label, which is similar to the labels already used in the European Union and the USA, presents complete information on a refrigerator's technical characteristics and daily and annual energy consumption. It also compares a given model with the best and the worst models of the same class produced in the Commonwealth of Independent States, and it estimates the consumer's annual energy bill with different tariffs for the worst and the best models available. "ENERGOCOMPASS" allows consumers to prepare for rising energy prices.

2.3. BUSINESSES

The mode of operation of industrial state-owned enterprises under a centrally planned system, with artificially low energy prices, gave no real incentives to energy managers to use energy efficiently. With the privatization of a great part of industrial enterprises, followed by the development of a private commercial and service sector, efficient use of energy as a universal production factor could become economically motivated. But general economic instability, the poor financial situation of the greater part of companies, and non-payments of energy bills, prevent energy saving from being considered a priority. Stabilizing production, developing markets for products, and establishing stable relations with business partners are considered by senior executive managers of companies to be the most urgent problems now.

Energy managers do not play an important role in decision-making in industrial enterprises. They are usually engineers who do not know not all that much about economics or system analysis. They are responsible mainly for the technical safety and reliability of the plant's energy system, and not for the efficient use of energy resources. They usually have no capabilities or incentives to implement energy saving measures.

At the same time, however, more and more examples of active interest in improving the energy efficiency of industrial production are observed among the larger industrial production companies. Usually an initiative comes from the senior executive managers (directors) who are coming to understand energy saving as a factor for improving the economic competitiveness of the products; this in turn leads energy managers and their services or departments to design and implement measures to save energy. The creation of an energy saving service at Magnitogorsky Metallurgical Plant is one of the most impressive examples (Nikiforov, G., 1998). Uralmash (the biggest heavy machinery manufacturer) has used energy efficiency projects as a key means to reduce its production costs (IEA/OECD, 1997).

3. Conclusion

A recent fundamental study on the dynamics of values in Russia in the course of structural and economic reforms (Lapin, 1996) only paid attention to health, freedom and other human values, but did not consider energy and its efficient use and related environmental issues as a basic value. Stabilization of the general economic and social situation in Russia is a basic precondition for all economic and social agents – private households, businesses, public services, administrations at all levels – to start to perceive energy saving and global climate change as significant criteria for decision-making and to realize rationales for relevant behaviour patterns.

To help people adjust to the new economic environment, changing their decision-making and behaviour in favour of more rational and efficient resource allocation – including energy resources – it is necessary to educate them in this field, starting from school, and explaining and demonstrating the benefits of energy efficiency with special reference to the implications for their health, security and other highly ranked values.

4. References

IEA/OECD, 1995: Energy Policies of the Russian Federation. OECD, Paris.

IEA/OECD, 1997: Energy Efficiency Initiative, vol. 2. IEA/OECD, Paris.

Lapin, N. (ed.), 1996: Dynamics of values in Russia in course of structural and economic reforms, Editorial URSS, Moscow.

Manning, N., O. Shkaratan, N. Tikhonova, 2000: Social and Employment Policies in Russia. Ash Gate Publishing Agreement, Hampshire.

Matrosov, Yu. A., 1997: Codes and Standards of Building Energy Efficiency: a Regional Approach. In: *Russian Energy Efficient Future: a Regional Approach*. Proceedings of the Conference at Chelyabinsk, RF, 25-26 September 1996, IEA, Paris, p. 231.

Nikiforov, G., 1998: The Assessment of the Efficiency of Power Consumption at Metallurgical Plant, Energomanager, no. 12.

RF Ministry of Fuel and Energy, 1998: Federal Target Program: Energy Saving in Russia. RF Ministry of Fuel and Energy. Moscow, 1998.

RF, 1999: Federal Law on Consumer Basket for the Whole RF. Moscow, November 1999.

UNEP, 1998: Climate Awareness Programme in Russia, Geneva, April 1998.

Vorobiev, V., V. Zhuze, S. Kuznetsov: Energy Audit Of Hospitals: Lessons For Future Energy Audits, Energy Efficiency, no. 19, pp. 8-11, April-June 1998.

List of Editors and Authors

Editors

Prof. Dr. Eberhard Jochem
Fraunhofer-Intitute for Systems and Innovation Research (ISI)
Breslauer Str. 48
76139 Karlsruhe, Germany
Phone: (+49) 721 6809 160
Fax: (+49) 721 6809 280
E-mail: Eberhard.Jochem@isi.fhg.de

Dr. Daniel Bouille
Piedras 482
2nd Floor Room H
1070 Buenos Aires, Argentina
Phone: (+54) 11 4331 1816/1649
Fax: (+54) 11 4334 4717
E-mail: ideefb@bariloche.com.ar

Dr. Jayant Sathaye
90/4000, Lawrence Berkeley National Laboratory (LBL)
1, Cyclotron Road
Berkeley CA 94720, USA
Phone: (+1) 510 486 6294
Fax: (+1) 510 486 6996
E-mail: jasathaye@lbl.gov

Authors

Dr. John A. "Skip" Laitner
Senior Economist for Technology Policy
EPA Office of Atmospheric Programs
1200 Pennsylvania Avenue NW, MS-6201J
Washington, DC 20460, USA
Phone: (+1) 202 564 9833
Fax: (+1) 202 565 2147
E-mail: Laitner.Skip@epamail.epa.gov

Dr. Stephen DeCanio,
Department of Economics
University of California
Santa Barbara CA 93106-9210, USA
Phone: (+1) 805 893 3130
Fax: (+1) 805 893 8830
E-mail: decanio@econ.ucsb.edu

Irene Peters
Swiss Federal Institute for Environmental Science and Technology (EAWAG)
Ueberlandstrasse 133
8600 Duebendorf, Switzerland
Phone: (+41) 1 823 5358
Fax: (+41) 1 823 5375
E-mail: ipeters@eawag.ch

Dr. Steven Ney
Interdisciplinary Centre for Comparative Research in the Social Sciences (ICCR)
Schottenfeldgasse 69, Tür 1
1070 Wien, Austria
Phone: (+43) 1 524 1393 160
Fax: (+43) 1 524 1393 200
E-mail: s.ney@iccr.co.at

Dr. Michael Thompson
The University of Bergen
LOS-centre, Rosenbergsgt. 39
5015 Bergen, Norway
Phone: (+47) 55 58 38 79
Fax: (+47) 55 58 39 01
E-mail: flomt@alf.uib.no

Dr. Harold Wilhite
Centre for Development and the Environment
University of Oslo
Postboks 1116 Blindern
0317 Oslo, Norway
Phone: (+47) 22 85 89 00
Fax: (+47) 22 85 89 20
E-mail: hal.wilhite@sum.uio.no

Dr. Elizabeth Shove
Department of Sociology
Cartmel College
Lancaster University
LA1 4YN, UK

Phone: (+44) 1524 594610
Fax: (+44) 1524 594256
E-mail:
E.Shove@lancaster.ac.uk

Prof. Loren Lutzenhiser
Associate Professor
Depts. of Sociology & Rural Sociology
Washington State University
Pullmann, WA 99164-4020, USA

Phone: (+1) 509/335 6707
Fax: (+1) 509/335 2125
E-mail: llutz@wsu.edu

Willet Kempton
College of Marine Studies, and Center for Energy and Environmental Policy
University of Delaware
Newark, DE 19716-3501, USA

Phone: (+1) 302 831 0049
Fax: (+1) 302 831 6838
E-Mail: willett@udel.edu
web:
www.ocean.udel.edu/faculty/willett/

Johanna Moisander
Helsinki School of Economics and Business Administration
Department of Marketing
P.O. Box 1210
00100 Helsinki, Finland

Phone: (+358) 943 131 (office)
(+358) 50 555 9935 (mobile)
Fax: (+358) 943 138 660
E-mail: moisande@hkkk.fi

Dr. Laurie Michaelis
Oxford Centre for the Environment
Ethics and Society
Mansfield College
Oxford, OX1 3TF, UK

Phone: (+44) 1865 282903
Fax: (+44) 1865 270886
E-mail: laurie.michaelis@mansfield.oxford.ac.uk

Dr. Minu Hemmati
UNED Forum
Co-facilitator of the CSD NGO Women's Caucus
3 Whitehall Court
London SW1A 2EL, UK

Phone: (+44) 171 839 1784
Fax: (+44) 171 930 5893
E-mail: minush@aol.com
Web: www.uned-uk.org
www.earthsummit2002.org/wssd

Prof. Richard B. Howarth
Environmental Studies Program
Dartmouth College
Hanover, New Hampshire 03755, USA

Phone: (+1) 603 646 2752
Fax: (+1) 603 646 1682
E-mail:
RBHowarth@Darmouth.edu

Dr. Stephan Ramesohl
Wuppertal Institute for Climate Environment Energy
PO Box 100480,
42004 Wuppertal, Germany

Phone: (+49) 202 2492 255
Fax: (+49) 202 2492 198)
E-mail:
stephan.ramesohl@wupperinst.org

Dr. Inna Gritsevich,
Center for Energy Efficiency (CENEf)
54, Korpus 4
Novocheremushkinskaya Strm
Moscow 117 418, Russia

Phone: (+7) 095 120 51 47/913 95 74
Fax: (+7) 095 883 95 63
E-mail: cenef@glasnet.ru

List of Reviewers, Rapporteurs, and Discussants

Dr. Altomonte, Hugo
Natural Resources and Infrastructure Division
United Nations, Economic Commission for Latin American and the Caribbean
UN-ECLAC
v. Dag Hammarskjold s/n, Vitacura
P.O. Box Casilla 179-D
Santiago, Chile
Phone: (+562) 210 2303; 210 2257; 10 2000
Fax: (+562) 208 0252
E-mail: haltomonte@eclac.cl

Dr. Annecke, Wendy
Energy and Development Research Centre
University of Cape Town
Private Bag Rondebosch
ZA-7701 South Africa
Phone: (+27) 21 650 2100
E-mail: wendy@energetic.uct.ac.za

Dr. Arimah, Ben
Department of Environmental Science
University of Botswana
P/B. 0022 Gaborone, Botswana
Phone: (+267) 355 3029
Fax: (+267) 585 097
E-mail: arimahbc@mopipi.ub.bw

Dr. Bonilla, David
Mechanical Systems Engineering
Tokyo University of Agriculture and Technology
2-24-16 Naka-machi, Koganei-shi
Tokyo 184-8588, Japan
Phone: (+81) 42 388 7076
E-mail: bonilla@star.cad.mech.tuat.ac.jp

Dr. Chanda, Raban
Department of Environmental Science
University of Botswana
Private Bag 0022
Gaborone, Botswana
Phone: (+267) 355 2521
Fax: (+267) 585 097
E-mail: chandar@mopipi.ub.bw

Prof. Crabbé, Philippe
University of Ottawa
Department of Economics
P.O. Box 450, Stn. A
Ottawa, Ontario Kin 6N5, Canada
Phone: (+1) 613 562 5874
Fax: (+1) 613 562 5873
E-mail: crabbe@uottawa.ca

Dr. Dang, Giap van
c/o S. Tejpar
1491 Nelson Avenue
West Vancouver, BC V7T 2G9, Canada
Phone: (+1) 604 925 6017
Fax: (+1) 604 925 6016
E-mail: vdang@ait.ac.th

Dr. Goldblatt, David
Centre for Energy Policy and Economics (CEPE)
ETHZ Zentrum WEC
8092 Zürich, Switzerland
Phone: (+41) 1 632 7520
Fax: (+41) 1 632 1050
E-Mail: david.goldblatt@cepe.mavt.ethz.ch

Dr. Heller, Tom
Institute for International Studies
School of Law
Stanford University
559 Nathan Abbott Way
Stanford, CA 94305-6055, USA
Phone: (+1) 650 723 7650
Fax: (+1) 650 725 0253
E-mail: theller@leland.stanford.edu

Prof. Kammen, Daniel M.
Energy and Resources Group (ERG)
310 Barrows Hall
University of California, Berkeley
CA 94720-3050, USA
Phone: (+1) 510 642 1139
Fax: (+1) 510 642 1085
E-mail: dkammen@socrates.berkeley.edu

Dr. Kühr, Helmut
DLR, German IPCC Co-ordination Office
Godesberger Allee 117
53175 Bonn, Germany
Phone: (+49) 228 4492 303
Fax: (+49) 228 4492 400
E-mail: helmut.kuehr@dlr.de

Dr. Molnar, Laszlo
Department for Innovation Studies and History of Technology
TU Budapest
Muegyetem rkp. 3
Budapest H-1111, Hungary
Phone: (+36) 1 463 1074
Fax: (+36) 1 463 1042
E-mail: molnarl@eik.bme.hu

Rooijen, Sascha van
ECN – Energy Research Foundation
Department: ECN - Policy Studies
P.O. Box 1
1755 ZG Petten, The Netherlands
Phone: (+31) 224 564143
Fax: (+31) 224 563338
E-mail: vanrooijen@ecn.nl

Dr. Sachs, Wolfgang
Wuppertal Institut
Abteilung Energie
Döppersberg 19
42103 Wuppertal, Germany
Phone: (+49) 202 2492-177
Fax: (+49) 202 2492-198
E-mail: wolfgang.sachs@wupperinst.org

Dr. Schaeffer, Roberto
Federal University of Rio de Janeiro
Graduate School of Engineering
Centro de Tecnologia - Sala C211
Caixa Postal 68565
21945-970 Rio de Janeiro – RJ, Brazil
Phone: (+55) 21 560 8995
Fax: (+55) 21 290 6626
E-mail: roberto@ppe.ufrj.br

Dr. Wamukonya, Njeri
Energy and Development Research Centre
University of Cape Town
Private Bag Rondebosch
ZA-7701, South Africa
Phone: (+27) 21 650 2100
E-mail: wamukonya@energetic.uct.ac.za

Dr. Xu, Huaqing
Energy Research Institute
State Development Planning Commission
Zhansimen Shahe
Beijing 102206, China
Phone: (+86) 10 697 331 06
Fax: (+86) 10 697 331 09
E-mail: xuhqing@public3.bta.net.cn

Dr. Yamba, Francis
Centre for Energy, Environment and Engineering (CEEEZ)
Great North Road
P.O. Box 32379 / Box E721 Plot 1635
Lusaka, Zambia

Phone: (+260) 126 2482 or (+260) 124 0267
Fax: (+260) 126 2482 or (+260) 124 026
E-mail: yamba@eng.unza.zm

Did not participate (excused)

Dr. Gupta, Sujata
TERI
Policy Analysis Division
India Habitat Centre
Lodi Road
New Delhi 110003, India

Phone: (+91) 11 462 2246
Fax: (+91) 11 462 1770 / 463 2609
E-mail: sujatag@teri.res.in

Dr. Jepma, Catrinus
University of Groningen
P.O. Box 800
8700 AV Groningen, The Netherlands

Phone: (+31) 50 363 3710
Fax: (+31) 50 363 3710
E-mail: c.j.jepma@eco.rug.nl

Prof. Lantermann, Ernst-Dieter
Gesamthochschule Kassel
FB 3
Holländische Str. 36
34127 Kassel, Germany

Phone: (+49) 561 804 3580
Fax: (+49) 561 804 3763
E-mail: lanter@hrz.uni-kassel.de

Dr. Rayner, Steve
Environment & Public Affairs
420 W. 118 St. IAB #1316
New York NY 10027
MC 3323, USA

Phone: (+1) 212 854 1688
Fax: (+1) 212 854 57566
E-mail: sr499@columbia.edu

Dr. Sokona, Jouba
Environment et developpement du tiers monde,
ENDA – Energie
54 Rue Carnot
BP 3370 Dakar , Senegal

Phone: (221) 8 225 983 / 222-496
Fax: (221) 8 217 595
E-mail: ysokona@enda.sn

Advances in Global Change Research

1. P. Martens and J. Rotmans (eds.): *Climate Change: An Integrated Perspective.* 1999 ISBN 0-7923-5996-8
2. A. Gillespie and W.C.G. Burns (eds.): *Climate Change in the South Pacific: Impacts and Responses in Australia, New Zealand, and Small Island States.* 2000 ISBN 0-7923-6077-X
3. J.L. Innes, M. Beniston and M.M. Verstraete (eds.): *Biomass Burning and Its Inter-Relationships with the Climate Systems.* 2000 ISBN 0-7923-6107-5
4. M.M. Verstraete, M. Menenti and J. Peltoniemi (eds.): *Observing Land from Space: Science, Customers and Technology.* 2000 ISBN 0-7923-6503-8
5. T. Skodvin: *Structure and Agent in the Scientific Diplomacy of Climate Change.* An Empirical Case Study of Science-Policy Interaction in the Intergovernmental Panel on Climate Change. 2000 ISBN 0-7923-6637-9
6. E. Jochem, J. Sathaye and D. Bouille (eds.): *Society, Behaviour, and Climate Change Mitigation.* 2000 ISBN 0-7923-6802-9

KLUWER ACADEMIC PUBLISHERS – DORDRECHT / BOSTON / LONDON